Issues in Bioethics
and the Concept of Scale

american
university
studies

Series V
Philosophy

Vol. 202

PETER LANG
New York • Washington, D.C./Baltimore • Bern
Frankfurt am Main • Berlin • Brussels • Vienna • Oxford

William A. Cook

Issues in Bioethics and the Concept of Scale

PETER LANG
New York • Washington, D.C./Baltimore • Bern
Frankfurt am Main • Berlin • Brussels • Vienna • Oxford

Library of Congress Cataloging-in-Publication Data

Cook, William A. (William Andrew).
Issues in bioethics and the concept of scale / William A. Cook.
p. cm. — (American university studies. Series V, Philosophy; v. 202)
Includes bibliographical references and index.
1. Bioethics. 2. Scaling (Social sciences). 3. System theory. I. Title. II. Series.
[DNLM: 1. Bioethical Issues. 2. Ecosystem. 3. Philosophy, Medical.
4. Systems Theory. WB 60 C771i 2009]
QH332.C676   174'.957—dc22   2008030362
ISBN 978-1-4331-0199-1
ISSN 0739-6392

Bibliographic information published by **Die Deutsche Bibliothek**.
Die Deutsche Bibliothek lists this publication in the "Deutsche
Nationalbibliografie"; detailed bibliographic data is available
on the Internet at http://dnb.ddb.de/.

© 2008 Peter Lang Publishing, Inc., New York
29 Broadway, 18th floor, New York, NY 10006
www.peterlang.com

All rights reserved.
Reprint or reproduction, even partially, in all forms such as microfilm,
xerography, microfiche, microcard, and offset strictly prohibited.

This work is dedicated to Jean, Willy, and David without whose love, understanding, and sacrifices I could never have proceeded with this endeavor. To them I give my love, gratitude, and apologies for their sacrifices.

TABLE OF CONTENTS

Acknowledgments   ix

Chapter 1   Bioethics   1

   Introduction   1

   The Nature of Bioethics and Bioethical Issues   3
      Bioethics   3
      Bioethical Issues   9

   Current Problems in Bioethics   14

Chapter 2   Systems Theory and the Concept of Scale   19

   An Introduction to Systems and Hierarchy Theory   19

   Hierarchy Theory, the Concept of Scale, and Bioethics   27

   Current Views of Scale   31
      Ecological Scales   31
      Societal Scales   32
      Correlative Ethics   37

   Proposed Framework of Scales   37

Chapter 3	**The Moral Nature of the Concept of Scale**	43

Description: Wholes and Parts   43

Prescription: Moral Decision-Making   51

    Scale and Guidance in Values Priority   53

    Human Nature and the Capacity of Foresight   59

    The Proposed Framework and Values Priority   60

Chapter 4	**Case Study: An Example of Scale**	69

Forestry and Clear-Cutting   69

    "Canada: A Model Forest Nation in the Making"   71

Recognizing Constraints   83

Discussion and Other Examples of Scale   96

    Resource Allocation in Medicine   96

    Global Health   101

    Maternal-Fetal Issues   106

    Sterilization of Mentally Incompetent Patients   109

    Regional Planning   112

Chapter 5	**Conclusions, Critique, and Reflection**	115

Central Contention   115

Supporting Arguments   116

    The Concept of Scale Helps in Considering Complex Systems   116

    The Concept of Scale has Prescriptive Implications for Values Prioritization   116

    Scales Are Evident in Bioethical Issues, and the Proposed Framework Helps to Elucidate Them   117

Adding Scale to Framework of Conflicts in Bioethical Issues   117

Critique and Reflection   118

References	123
Index	131

ACKNOWLEDGMENTS

I would like to acknowledge the University of New Brunswick School of Graduate Studies for its leading role in the development of post-graduate interdisciplinary studies; my Graduate Committee, Drs. Norm Whitney, Alan Miller, Brian Cupples, Warren Neill and Tilman Benfy, for their constant encouragement, critique, and understanding; my many friends and colleagues whose patient listening and subsequent commentary and support were always desired and appreciated, especially Jean Cook, Bill Randall, Don MacDougall, John Coates, John Gillis, Eric Gozna, Karen Watson, Deb Dickeson, Philip Beck, Bev Schneider and Wendy Carty; my medical colleagues who patiently put up with me and my study requirements; my editor, Heidi Burns without whose interest, assistance and encouragement this book would not have happened; and finally, my parents, Eva and Bill, whose never-ending love and encouragement have shaped me and guided me to this work.

· 1 ·
BIOETHICS

Introduction

In 1973, when I was a clinical clerk (a final-year medical student), I spent two months on the Queen Charlotte Islands doing an elective in Family Practice. I was based in Queen Charlotte City at the small community hospital and clinic that served all of the Queen Charlottes. My responsibilities included attending health clinics in outlying communities including several logging camps in relatively inaccessible locations. My interests at this particular time in my life were focused on my family (my wife and I were expecting our first child) and on completing my medical training. However, strong as these interests were, I was overwhelmed by the vistas that were before me as I traveled the Queen Charlottes by foot, auto, boat, and plane. To this day, I have vivid memories of the beauty of the land, forests, rivers, and sea. I also have vivid memories of the destruction and ruin of the landscape left by the lumber industry. Large tracts of clear-cut forests with fresh and old stumps and chaff were evident everywhere I went. They were present even on steep hillsides and mountain slopes where the wind and rains (the Queen Charlottes are a temperate rain forest) had washed the soil into the rivers and ocean. Intuitively, I felt and knew that this was inappropriate and could not understand why or how such destruction could be allowed. I thought that there must be a better way to harvest the forest resources without leaving such apparent chaos and wondered what was in the minds of the lumber company officials and the government that they could not see what they were doing. These questions still haunt me today and have structured my reading and thinking as much as has my medical practice. I have come to understand that human use of natural resources is as much an ethical endeavor as is the practice of medicine, and that considerations beyond the local and economic scales are critically important.

Since 1988, I have been heavily involved in the study of bioethics and have been an active participant in this field within my professional medical community. I have overseen the development of an Institutional Ethics Committee (IEC) for the Health Region in which I practice. Also, I carry on a Bioethics Service which functions as the service arm of the IEC. Through the Bioethics Service, I have undertaken regular educational rounds, seminars and in-services, and clinical ethics consultations.

Drawing on these experiences, I have come to the conclusion that it is crucially important to address our responsibilities to the natural environment. Furthermore, I believe that addressing these responsibilities in a particular manner will illuminate, if not solve, certain dilemmas in the field of bioethics. Specifically, I argue that the concept of scale, as I abstract it from hierarchy theory as applied to ecological systems, is an important notion for clarifying conflicts not only in regard to environmental issues in bioethics, but in other areas of concern to bioethics as well. Accordingly, I intend to argue here that this concept of scale, as understood from hierarchy theory used in ecology, is helpful in identifying scalar sources of conflict, i.e., conflict among the hierarchy of systems (elaborated in chapter 2). I also argue that this concept of scale is of value as an aid to discernment among values and, thusly, in choosing our course of action in many bioethical issues. A practical means for addressing scale and incorporating it into bioethical analysis and decision-making is the focus of this book. This work is not intended to be an argument in moral philosophy but rather an attempt to integrate a concept from ecology (a feeder discipline to bioethics) into the framework of issues in bioethics promoted by Roy, Williams and Dickens (1994) in their book *Bioethics in Canada*. My objectives are to clarify:

- the nature of bioethics;
- the nature of scale as I intend it;
- the moral nature of this concept of scale;
- how scale is presently considered;

and to present

- a conceptual framework for considering scale in bioethics.

The concept of scale from ecological hierarchy theory has been found useful in dealing with the varied and complex nature of relationships and interrelationships in natural systems and, as I will demonstrate, has a moral dimension that may be used as a normative guide to action in bioethics. I wish to emphasize that I am concerned here with a conceptual approach to considering issues in bioethics. I frequently refer to humanity or humankind using terms like "we" and "our," and I will clearly note when I refer to specific individuals or groups rather than to humanity, or human moral agents in general.

The Nature of Bioethics and Bioethical Issues

Bioethics

Since the term *bioethics* was coined, it has become a discipline in its own right. It has its own literature, tradition, and history of development and thought. It has its own largely agreed upon and continually developing subject matter and methodology that are distinct from other branches of ethics, philosophy, medicine, biology, law, nursing, social work, epidemiology, and ecology, all of which are potential feeder disciplines. In other words, bioethics has its own practitioners who come from these varying backgrounds and perspectives to service a recognized constituency. Bioethics, as a discipline, represents a move away from a preoccupation with pure ethical theory to a renewal of an Aristotelian approach giving greater attention to practical reasoning (Roy et al., 1994). Roy et al. (1994), also, clearly distinguish bioethics from morality (by which they mean an expressed set of prioritized values); from philosophical ethics (by which they are referring to metaethics, normative ethics, and applied ethics). Briefly, and in simplified form, I understand Roy et al.'s (1994, pp. 37–58) argument for this distinction to be focused on the following points. First, the nature of bioethics is not that of a morality with an expressed set of prioritized values. Second, bioethics is not a normative ethic, but it does use normative systems in its process of practical reasoning to reach various practical judgments about what is permissible, what is tolerable, what is mandatory, and what is prohibited. Third, bioethics is not concerned with critique or analysis of the concepts of ethics and ethical theories (the realm of metaethics) but may make use of notions from metaethics in its practical reasoning. Fourth, bioethics is not a type of applied ethics because primary among other reasons, there is no preset ethical theory or normative order to be applied. The method of bioethics is not dedicated to a deductive application of explicate principles and rules to specific cases and issues. Rather, the order of values, principles, rules, and reasons is implicate in nature, being discovered or uncovered as the particularities of the situation unfold and become known. I take this description of bioethics to be valid primarily on the basis of my experience over ten years in both clinical ethics and on institutional ethics committees where the cogency of the reasons noted above has been evident to me.

Bioethics has been defined by Gorovitz (1977) as "the critical examination of the moral dimension of decision-making in health related contexts and in contexts involving the biological sciences." Engelhardt (1988) defined bioethics as "the disciplined analysis of the moral and conceptual assumptions of medicine, the biomedical sciences, and the allied health professions." Excellent historical reviews of the development of bioethics as an independent discipline can be found in Roy et al. (1994), Reich (1994), and Katz (1994). In a personal account of the development of bioethics, Katz (1994) described how, in 1964, while auditing a law class at Yale where he was on faculty, a discussion about protecting an accused person from the forces of the state led him to contemplate his own earlier involvement as a medical researcher in dream research. In this research subjects where given hypnotic suggestions regarding the content of their dreams. Katz and his fellow researchers actually considered the possibility that some psychological harm to the subjects could come of this, so they arranged for their institution to provide subjects with free care, if the need should arise. What struck Katz during the law class was that it never really dawned on him or his research colleagues that they should disclose their concern to the research subjects, or even not to proceed with the studies. This experience led him to read the transcripts of the Nuremberg trials, recounting the atrocities of medical experiments carried out on human beings in the Nazi concentration camps before and during World War II. Such experiments were ostensibly for the benefit of Aryan peoples and were not only without benefit, but were directly harmful to the subjects, not to mention being undertaken without their consent. Katz then proceeded to look into the literature on human experimentation only to find it extremely scant with very little attention paid to the protection of the research subject. At this point, Katz realized that there was much work and study to be done to safeguard subjects in research projects, and he became active in this regard, finding that there were a few other lonely souls who had come to the same conclusion. During the 1965–66 academic year, Katz started a seminar class at Yale Law School entitled "Experimentation on People." The class was geared to look specifically at the issues "of consent and the role of the state in the supervision of medical experimentation" (Katz, 1994, pp. 87–88) on people.

At about the same time, Beecher, an anaesthesiologist and clinical researcher, published his now renowned exposé that might be considered the inception of Research Ethics, a field of bioethics concerned specifically with ethical issues (particularly the welfare of research subjects) in research in the life sciences. His concern was with the failure of researchers to inform their subjects adequately, not only of the risks of the experiment for which they had "volunteered," but also that they were actually subjects in an experiment, even when grave consequences could be expected (Beecher, 1966). All this activity stimulated much discussion in academic circles in the United States about the ethics of human experimentation and, in the late 1970s, culminated in the President's National Commission for the Protection of Human Subjects of Biomedical and Behavioral Research. The resulting Belmont Report (U.S., National Commission, 1978) laid out a framework of principles that seemed to be pertinent and applicable to these concerns about ethics and research. Interestingly, this process revealed that the participants were frequently at loggerheads about ethical theory and justification for certain beliefs and actions. However, it became clear to them that they could come to an agreement on a set of general principles that seemed to capture the respect and honor owed to persons and thusly provided tenets for good and right behavior when involving people in research. These principles included respecting the subjects as autonomous persons and recognizing that a fuller disclosure of information (part of what later became known as informed consent) was a necessary part of participation in clinical research. This support for individual autonomy was a new phenomenon in medicine (which had typically had a paternalistic approach to patient care), and has been a dominating focus in the emerging field of bioethics, especially in the United States ever since. In addition, the long recognized principles of avoiding harming others (nonmaleficience) and not only protecting them from harm but promoting their welfare (beneficence) were readily supported as common interests for good and right behavior in research. Similarly, consideration of fairness (justice) was also found to be a common concern.

These four principles, respect for autonomy, beneficence, nonmaleficence, and justice, became a core framework for analysis in the emerging field of bioethics and were subsequently championed by Beauchamp and Childress at The Kennedy Institute for Ethics founded in 1971 at Georgetown University in Washington, DC. This "principlism" approach to issues in bioethics (Beauchamp & Childress, 1994), almost exclusively clinical ethics, has been a leading force in bioethical analysis and decision-making. However, this approach has consistently been challenged by some who feel that overemphasis on principles potentially creates a formula-based calculation with autonomy consistently trumping the other principles (Toulmin, 1981). This, then, leaves relatively unconsidered the key role played by virtues, virtue ethics (Beauchamp & Childress, 1994; MacIntyre, 1984; Pellagrino, 1985), moral emotions, and responsiveness in relationships, care ethic (Baier, 1985; Beauchamp & Childress, 1994; Blum, 1988; Carse, 1991; Gilligan, 1982), and our experience with other similar circumstances, casuistry (Arras, 1991; Beauchamp & Childress, 1994; Jonsen & Toulmin, 1988), in our choices, behavior, and relationships. These differing approaches suggest that a principles approach to bioethical choices, by itself, misses the depth and richness of human context and is therefore incomplete.

At about the same time that these issues in research ethics were coming to the fore, a medical oncologist from Madison, Wisconsin, Van Rennselaer Potter, was writing and speaking about his concerns for the survival of the human species. He was prominent in drawing attention to our relationships, not only with each other, but with the natural world in which we live. Potter's concern lay with sustaining a healthy, long-term relationship between humanity and the natural environment. It is Potter who coined the term *bioethics* in 1970–71 and advocated it as "a new discipline that combines biological knowledge with a knowledge of human value systems" (Potter, 1971). He explained, subsequently, that he "chose 'bio' to represent biological knowledge, the science of living systems; and ... 'ethics' to represent knowledge of human value systems" (Reich, 1994, p. 321). This conception of bioethics is clearly much broader in scope than the field being developed in medicine and its immediate biological cousins (genetics, molecular biology, and so on). Indeed, the broader areas of concern that Potter advocated, dealing with human survival in the natural context, expanded into the field of Environmental Ethics. I agree with Potter's view that bioethics should be broad in its scope, and there are now signs of growing support for this position. For instance, the concept of scale in ecology and environmental planning has been taken up by Norton (1994b) while Roy et al. (1994) provided a commentary on the ethical aspects of current environmental threats to human life. Roy et al. argued that this challenge to ecological integrity, coming from unwise human behavior, is promoting a growing interest in environmental and ecological ethics. They note that environmental concerns have been present throughout the duration of the development of bioethics as a discipline and comment that "now, however, it can no longer be ignored" (p. 468).

In addition to the influence of these environmental concerns on bioethics, rapid advances in medical knowledge from the 1950s onward not only challenged our views of life and death but also have generated an intense debate about what medicine can and should do. Thus, we see advances in transplantation techniques have stimulated debate on a variety of ethical issues, such as how to allocate organs in short supply; who should pay; and how resources should be fairly distributed given that benefits to a few may reduce resources to others. Similarly, the development of our capabilities to resuscitate people from death (CPR) and sustain life (heart and lung function) have led to renewed discussions on what constitutes a normal or acceptable human life and when, if ever, it might be appropriate and morally sound to withhold or withdraw such interventions. Since these debates are about matters of life and death, it did not take long before the public was demanding answers and the courts (especially in the United States) began hearing these issues. The subsequent common law precedents were of great moral impact for the developing field of bioethics. Two such cases from the United States, the 1973 *Roe v. Wade* case that saw the U.S. Supreme Court set a legal standard for abortion, and the 1976 case of Karen Ann Quinlan that saw two courts in New Jersey set standards for considering the withdrawal of treatment from a permanently unconscious patient, are good examples of this discussion. Both these cases set the tone for subsequent ethical debate underlying the medical choices involved in such dilemmas as abortion and withdrawing life-sustaining treatments.

As outlined above, Potter coined the term *bioethics* for a new discipline concerned with understanding human values and human well-being in the context of the biomedical and life sciences. Roy et al. (1994), as explained earlier, characterized bioethics as distinct from philosophical ethics and morality and focused on practical reasoning around issues of moral concern in the various fields of the life sciences. For the purposes of this book, I adopt the operational understanding of bioethics that reflects an emphasis on practical reasoning and Potter's (1988) concern with the need for an integration of "short-term views" (characteristic, in his view, of biomedical ethics) and "long-term views" (characteristic, in his view, of environmental ethics) (p. 74). Potter makes it quite clear that "the time has come to recognize that we can no longer examine medical options without considering ecological science and the larger problems of society on a global scale" (1988, p. 2). Engelhardt, in his introduction to Potter's *Global Bioethics* (1988), emphasized Potter's view stating, "we must reflect on the values ingredient in our relationship to nature if we are to secure our continued survival in a way that will accord with our views of our own well-being ... we will need to understand what kinds of values concerning the environment impose what kinds of obligations" (1988, pp. x–xi). These comments receive support from mounting evidence that the recent outbreaks of new and deadly infectious diseases are largely a result of global environmental destruction and disruption of both terrestrial and marine ecosystems (Kain, 1995; Last, 1993; Loevinsohn, 1994; McMichael, 1993). In his comments on the metaphors of medicine, I believe that Annas (1995) supports Potter's view when he suggests that medicine needs to take on an ecological worldview. Further support comes from what may be perceived as a general intuitive awareness that human health is directly related to the health of the environment within which we live. Rising skin cancer rates make us think of depleting ozone levels; increasing incidence of, and deaths from, asthma turn our thoughts to air quality; and malnourishment and starvation cause us to reflect upon droughts and crop failures, which, in large part, are a consequence of excessive human exploitation of natural systems.

Potter did not give a clear description, however, of how an integration of values and facts, short- and long-term views, or medical and environmental concerns, could be accomplished. Furthermore, Western systems of moral thought have, for the most part, primary concern for inter-human, interpersonal relationships within a long-standing worldview that sees humans as separate from, and generally in conflict with, nature, by which I mean that natural limitations and constraints are viewed as obstacles to be overcome or, from the view of economics, as externalities that do not rate consideration. For this reason, present systems have experienced difficulty dealing with ethical concerns beyond the strictly human interpersonal level. The concept of the "tragedy of the commons" (Hardin, 1968) presents an example of what I mean. The classic description of this scenario involves a pasture held in common by all people accessing it. Each herdsman rationally adds another head of cattle to his herd grazing on the commons since the benefit to him is his alone while the disbenefit (harm) of overgrazing is shared by all. The limitations of the pasture (the ecosystemic carrying capacity) are ignored and, eventually, "freedom in the commons brings ruin to all" (Hardin, 1968). This situation might be initially approached as a technical problem and, with the addition of nutrients to the pasture, in the form of fertilizers and perhaps even genetically engineered grasses, the initial limitations of the natural ecosystem may be overcome for a period of time. However, this remedy will not last as the tragedy recurs in an unending spiral. As Hardin (1968) points out, this is an example of a problem with no technical solution and his stated solution to this is to understand the necessity of "mutual coercion, mutually agreed upon." This description does seem to be the nature of societal constraints represented by laws and regulations. This scenario can be seen playing out to its foreseeable conclusion in such situations as the loss of cod stocks on the Grand Banks of Canada, and in the over-logging occurring on woodlots in various locales in Canada. Hardin (1968) further points out that the goodness or badness of an act is "system-sensitive," that is, it is contingent upon the state of the system at the time the act occurs. Hardin (1968) notes, rightly or wrongly, that this system-sensitivity had been overlooked by most previous moral philosophers. In my view, from my reading and experience, present human ethical systems struggle with the notion of giving moral weight to the limitations and constraints of the greater whole of which we are a part. The need for a broader bio-ecological approach (biological in the sense of all life and ecological in the sense of all biotic and abiotic relationships) has been recognized by a variety of scholars (Hugo, 1890/1992; Leopold, 1949/1970; Odum, 1989; Schweitzer, 1949). Roy et al. (1994) consider these environmental and ecological ethical issues to be "integral dimensions of bioethics" (p. 21) and see this realm as an area for future development in bioethics. More specifically, bioethics deals with ethical concerns in the life sciences, and it is now recognized how interconnected human life is with other natural systems. The discussion of matters of scale offered in this book is one small step toward the development of a broader basis for bioethics.

Bioethical Issues

Roy et al. (1994) have also characterized bioethics as "issue oriented" (p. 29). An issue is seen as involving "conflict and controversy" (p. 30) over what is mandatory, permissible, tolerable, or prohibited in human behavior. They state:

> Since the way to resolve one kind of issue will often not work for issues involving conflicts of quite different kinds, it is essential to recognize precisely what kind of conflict we are dealing with, when people strongly disagree about what should or should not be done in medicine and in applications of the life sciences and technologies to human life. (p. 30)

This is just as true, as well, for human activities with regard to our biological and ecological relationships. "An issue, then, involves conflict and controversy. The two fundamental questions of bioethics, the *what* [substance] and the *how* [process] questions, return over and over again. ... Because the answers to these questions are so often diverse and conflicting, the questions become or generate issues" (Roy et al., 1994, p. 30). However, as Roy et al. point out, issues are not all of the same nature. Issues differ from one another in terms of the conflict(s) that is (are) involved. To recognize precisely what kind of conflict one is dealing with, Roy et al. (1994, pp. 30–37) offer a framework consisting of seven categories of conflicts. These are:

1. conflicts based on insufficient evidence and experience;
2. conflicts based on partial perceptions;
3. conflicts arising from a restricted value focus;
4. conflicts arising from unexamined assumptions of fact and instrumental beliefs;
5. conflicts at the level of ethos (world-view beliefs);
6. conflicts on the level of morality (values and principles);
7. conflicts and confusions on the level of ethics.

Conflicts based on insufficient evidence and experience are essentially conflicts over facts and data. This dynamic is not unique to bioethics but is common to all fields of inquiry. When information is incomplete or unavailable, one's decisions may be skewed by the lack of data to provide guidance in deciding what to do. For example, a patient may have to choose between therapeutic interventions but has trouble making the choice because of uncertainties in the risks and benefits. This indecisiveness is particularly common in cases where treatments are new or experimental, for example chemotherapy for cancer, and our experience with them is limited. Another example from a public policy perspective is the controversy around genetic engineering and recombinant DNA technology. When these techniques were developed in the early 1970s, many were eager to explore them on the assumption that risks were small, while others were concerned about the lack of knowledge about the effects they might have. There were no experience and no data to help decide what to do. Scientists met at Asilomar, California, in 1974 to develop guidelines for this type of research at least until more experience and knowledge about outcomes could be gained. Roy et al. have suggested the following strategy for dealing with conflicts of this type:

> When the social consequences are difficult to foresee and are uncertain, new developments need a trial run under close monitoring until a sufficient accumulation of knowledge and experience reveals whether limits, restrictions, and protective barriers have to be established. It is difficult to judge new developments adequately by using thought experiments alone. (p. 31, original emphasis)

Conflicts based on partial perceptions are a broad category implying that the focus of any one perception of an issue may be so narrow that it misses the greater picture, as in not seeing the forest for the trees, as it were. In this circumstance, disagreement arises between protagonists, all of whom believe that they are viewing the whole picture when, in fact, they are seeing only a part of it, i.e., taking a part to be the whole. I frequently see an example of this type of conflict in my clinical ethics rounds in the neonatal and labor and delivery units. Here, the nurses are geared to the care of, and concern for, the well-being of the fetus. They find it very difficult to understand the behavior of a pregnant woman, admitted for management of illness associated with drug use, who seems to hold the well-being of the fetus in much lower regard. "Why can't this patient be forced to care for the fetus appropriately?" they ask. It is easy to lose sight of the fact that the pregnant woman is the patient and that she may well have many other aspects to her story that the single-minded concern for the fetus does not illuminate. For these types of conflicts over partial perceptions and limited information, Roy et al. have offered this strategy:

> Conflicting positions based on partial perceptions and limited information can usually be resolved when the parties in conflict can be brought to accept information about the total phenomenon; and when they can come to adopt a higher viewpoint that will integrate and balance the partial perceptions. (p. 31, original emphasis)

Conflicts arising from a restricted value-focus are very common and result, as this classification suggests, from an overemphasis, or a singular emphasis, on one value when another, or many values, may be at play. A classic example of this situation is the issue of abortion in which we detect confrontational, adversarial positions from two main groups supporting their value foci, namely, "pro-life" and "pro-choice." Similar confrontational and adversarial positions can be seen in the issues around preservation of the environment. Here, at the polar ends of the spectrum of concern for the environment, "environmentalists" and "industrialists" promote bio-centric and anthropocentric values, respectively. Again, Roy et al. have suggested a strategy to this type of conflict:

> An issue rooted in conflicts based upon a restricted value-focus can hardly ever be resolved unless each party to the conflict acquires an understanding of, and respect for, the values the other party is trying to protect. That is unlikely to happen without interdisciplinary, interprofessional and intercultural collaboration. (p. 32, original emphasis)

Another category of conflict in bioethics has to do with what might arise from the failure to test critically our underlying assumptions about a thing. These assumptions are notions we accept as being true and accurate descriptions of the way things are. In general, it is expedient to take such assumptions for granted without routinely evaluating them, otherwise getting along in life would be cumbersome.

However, our assumptions are not always as accurate as we think they are, and it behooves us, on occasion, to be critical of them, to reassess them and, when indicated, to amend them. A classic example of an unexamined assumption of fact is found in the story of the world explorer who happens upon a tribe of cannibals. After some discussion, the explorer is satisfied that this tribe, by their own account, does not eat people, and he accepted their invitation to dinner. He soon discovered his mistake, as it became clear that, indeed, he was to be dinner. "I thought you don't eat people," he said, only to have it finally clarified that they don't eat people, but since he is not a member of the tribe, he is not "people."

Assumptions of fact are not the only assumptions that get us in trouble. As individuals, we have extensive knowledge about the way things work in our everyday life. Most of this knowledge, however, is not gleaned from personal experience, but from our willingness to accept, as true, conventional wisdom. Such beliefs play an instrumental role in that they allow us to carry on with our lives without undue demands for new information and skills. Conflicts in bioethics may develop when such assumed beliefs suggest different, often conflicting, ways of responding to the situation at hand. An example of this type of conflict is evident in the issue of providing intravenous (IV) hydration to dying patients. Most people assume that water is a necessary requirement of life and that not providing hydration would create major suffering for patients. It has, therefore, been felt to be unethical to withhold fluids from dying patients because it would increase their suffering. Certainly, withholding fluids from healthy individuals does result in an identifiable and measurable suffering, but is this also the case with terminally ill patients? New evidence suggests that the very ill and dying do not experience dehydration as do the healthy, and that hydration may actually cause suffering by keeping mucous production at a level that results in congestion and the inability to clear the chest and throat. A reassessment of our beliefs about hydrating the dying has been necessary.

Roy et al. (1994) have observed that conflicts over facts and instrumental beliefs are not resolvable by pure argument, frequently necessitating the introduction of new information through research especially. The process here requires "*the unmasking of the mistake in one or the other or in both beliefs. These mistakes may not be easily detectable by simple observation, and their unmasking may well require well-organized research*" (p. 34, original emphasis). Resolving conflicts in this category therefore requires the explicit questioning of the assumption and usually new data and evidence to illuminate it.

Conflict also occurs at the level of fundamental beliefs, or worldviews. These are paradigmatic understandings and assumptions of the structure, function and meaning of our lives as individuals and as a society. Roy et al. (1994) refer to a worldview as an ethos, "governing beliefs we have about the status, destiny, and meaning of human life, or about the order of society and the purposes and roles of societal institutions in maintaining human community" (p. 34). Bioethics is ripe with conflicts of this type where such fundamental beliefs often clash. Euthanasia and physician-assisted death, abortion, genetic engineering, national health care, the use of our natural resources, and the preservation and restoration of wilderness, are all examples of issues where underlying worldviews are in conflict. Since worldview beliefs "are the field, the lighted space, or the horizon within which data, facts, and information are interpreted and acquire their meaning and their value" (Roy et al., 1994, p. 34) they are not readily altered by new data, facts, or information. Roy et al. offered the following advice for resolving conflicts over ethos:

> *The resolution, if ever attainable, requires, on the part of one or many groups or individuals, a change of belief so profound that the change involves a personal or community conversion. In these situations, the best attainable solution is often a political accommodation ... that maintains the coherence of a society, fosters the process of moral and philosophical discourse and debate, supports respect for personal conscience, and protects minorities (whether they be religious, cultural or moral) from subjection to the dictates of majorities without justifying cause.* (p. 35, original emphasis)

Agreeing to disagree and reaching some sort of compromise that all can live with is often the pragmatic outcome in this type of conflict.

In our moral lives, we choose our behavior based on a combination of habit, response to authority, perceived consequences, and sets of values and principles, all of which find foundation in our underlying ethos. Without arguing about or confronting our worldviews, we may simply argue about the principles or value sets to which we subscribe since we often differ over the ordering of these principles and values. Roy et al. refer to this category of conflict as "conflict on the level of morality" and characterize it as "a conflict about which values may be sacrificed and which may not, when two or more values cannot be honored or achieved in a given situation" (p. 35). The moral dilemma of abortion is such an example. How should the value of caring for the fetus (as a potential new human life) be reconciled with the other values that the mother involved holds dear (these may include responsibility and caring for self and for other children, and a life plan with goals and aspirations)? Our view of the natural world is replete with economic, ecological, recreational, aesthetic, and spiritual values, to name a few, and most everyone is aware of the conflicts among them. In attempts to reconcile conflicts over recognized principles, Roy et al. (1994) have suggested a "*rigorous clarification of arguments about the foundation, implications, scope, and interrelationship of principles*" (p. 36, original emphasis) within the context of the particular circumstance at hand, specifically in clinical ethics, the "*body and biography*" (p. 36, original emphasis) of the patient. They recognize that conflicts rooted in value priorities are more difficult to resolve and suggested that:

> In these morality-conflict situations, the most practical and civilized strategy aims at a compromise. A policy is sought which sets the conditions under which a morally disputed behaviour will be at least socially tolerated by most people, even if it may not be socially championed by many. (p. 36, original emphasis)

Roy et al.'s final category of conflict revolves around the confusion they see about "bioethics as ethics," that is, the conflict and confusion over what constitutes bioethics and its method, and how it relates to ethics in general. This category of conflict is stated here to recognize it only. Like the work of Roy et al., this book is not directed toward discussion of philosophical ethics (metaethics, normative ethics, applied ethics) or morality, but toward the inclusion of the notion of scale, as understood from an ecological perspective, as a source of conflict in bioethical issues, and in decision-making in these issues.

I believe that the framework for considering bioethical issues elaborated by Roy et al. (1994), and described in the preceding paragraphs, is useful in clarifying the precise nature of the conflict inherent in the issue. This is important because each type of issue requires a potentially different approach for resolution. However, as indicated above, I think that this framework is incomplete because it does not explicitly address scale as a possible source of conflict in bioethics. The scalar nature of activities, relationships, and values is a core concept in Hierarchy Theory as applied and elaborated in ecology (Allen & Starr, 1982; Holling, 1992; O'Neill et al., 1986). This concept can be used in bioethical analysis, synthesis, and decision-making. Bioethical analysis implies critical reflection and consideration to achieve conceptual clarity, and synthesis encompasses the dialogue that carries the analysis through its own process of questioning and testing with a resultant decision as outcome.

Current Problems in Bioethics

> In the relations of humans with the animals, with the flowers, with the objects of creation, there is a whole great ethic scarcely seen as yet, but which will eventually

break through into the light and be the corollary and the complement to human ethics. ... Doubtless it was first of all necessary to civilize man in relation to his fellow men ... but it is also necessary to civilize humans in relation to nature. There, everything remains to be done. (Victor Hugo, 1890/1992)

In this passage from his diary, Victor Hugo recognizes a struggle, a challenge that is weighing heavily on humanity today. Roy et al. (1994) have recognized this challenge in the introductory chapter of their text *Bioethics in Canada* (p. 21). Any brief review of the literature in bioethics will show that our present approaches to bioethics are having some difficulty in coming to grips with many of our contemporary dilemmas. There is an abundance of literature in many disciplines on interhuman ethics in a pluralistic society (Beauchamp & Childress, 1994; Goodin, 1980/1987; Levine, 1993; Midgley, 1983/1987; Narveson, 1980/1987; Roy et al., 1994) and on humanity's relationship with the natural world (Callahan, 1994; Capra, 1982; Cheney, 1989; Devall & Sessions, 1985; Frye, 1982/1987; Gardiner, 1990; Midgley, 1983/1987; Milbrath, 1989; Norton, 1991; Potter, 1988; Rolston, 1988; Sahtouris, 1989). Similarly, much has been written on environmental issues and the difficulties in policy development with respect to human use and abuse of the environment (Callicot, 1986, 1992; Coufal & Spuches, 1995; Ehrenfeld, 1994; Milbrath, 1989; Norton, 1991; Odum, 1989; Sagoff, 1992). Similarly, one can find discussions on intergenerational ethics and the interrelationship of environmental issues and future generations of humankind (Brook, 1980/1987; Goodin, 1980/1987; Green, 1977; Narveson, 1980/1987; Norton, 1989, 1992; Splash, 1993), and on interspecies or interspecific ethics with regard to animal rights especially, but others as well (Goodin, 1980/1987; Midgley, 1983/1987; Rawls, 1972; Singer, 1975, 1986; Stone, 1974). Most texts and journals on ethics will have two, if not more, differing ethical views on these issues and offer their approach to dealing with them. However, there often remains no satisfactory way of choosing between or among the differing views or methods (Beauchamp & Childress, 1994; Cragg, 1987; Jonsen et al., 1992; Levine, 1993; Potter, 1975, 1988; Regan, 1986; Rolston, 1988; Roy et al., 1994; Splash, 1993). These challenges to our ethical systems and the suggestion that something more or some other approach to them should be considered was forecast by several prominent individuals (Hugo, 1890; Leopold, 1949/1970; Schweitzer, 1949).

Midgley (1983/1987) recognizes these issues and identifies broader relationships which, she suggests, carry duties and responsibilities that our present ethical systems have difficulty with. Midgley comments that "it is remarkable that we now have to state this obvious truth as if it were new, and invent the word 'ecological' to describe a whole vast class of duties" (p. 444). Allen and Starr (1982) suggest that "with the gradual encroachment of ecology into larger and smaller realms... ecological insights do begin to be won outside the commonplace human experience" (p. 4).

In his book *Global Bioethics*, Potter (1988) suggests that C. H. Waddington (a well known naturalist, geneticist, and paleontologist of the first half of the twentieth century) was the first bioethicist, as he expressed the view that our human ethical systems and theories needed to take into account our biological knowledge. "What is demanded of each generation is a theory of ethics which is neither a mere rationalization of prejudices, nor a philosophical discourse so abstract as to be irrelevant to the practical problems with which mankind is faced at that time" (Waddington, 1967, p. 19). Particularly germane in our present era are the unprecedented environmental (overpopulation, resource depletion, pollution, loss of biodiversity), medical (resource allocation, resurgence of ancient diseases, new diseases) and technoscientific advances (genetic engineering) that threaten to overwhelm us, as well as throwing us into much confusion. The common denominator in these problem areas is often the perceived conflict between either the needs of the individual and those of the human collective or individual and collective human needs, on the one hand, and those of the natural environment, on the other. Such conflict between values is the stuff of ethics, and it is precisely these kinds of problems with which our current ethical systems are having difficulty in coping. When considering problems like these using the framework outlined by Roy et al. (1994), we recognize conflicts not only about values (morality), but also over worldviews (ethos), partial perceptions, and a restricted value-focus. One explanation of why our current ethical systems are having trouble dealing with these conflicts is that there is a strong tendency toward polarization. More precisely, analysis and resolution are often seen as an either/or situation. For example, the argument about logging is commonly stated as one about protecting and promoting jobs or protecting and promoting ecosystem biodiversity but not as protecting and promoting both. The problem is that present approaches fail to consider parts and wholes in a way that captures the true nature of life, that is, that everything is integrated in a complex web of interdependent relationships. There are issues that are amenable to analysis at the level of the parts, for example, between a patient and her doctor. In addition, there are issues that require analysis at the level of the whole where the interrelationships and interdependencies are critically important, for example, between patients and public health medicine, or between loggers and the forest ecosystem. The consideration of parts and wholes, following the notion of scale as extracted from ecological hierarchy theory, may be a useful adjunct to analysis and decision-making in bioethics that helps in identifying and sorting out the nature of the conflicts involved.

Subsequently, I will consider some bioethical problems through which I hope to show that there is an important class of conflicts in bioethical issues over and above the seven classes of conflict outlined by Roy et al. (1994). Furthermore, I will show that this other way of viewing the issues (from the perspective of scale) will help in analyzing them and also provide guidance in resolving them. Since bioethics is concerned with issues of human health and the environment, as discussed earlier, and since the aspect of scale addressed in this book comes from ecological science, it is fitting to use, as the primary case study, an environmental issue in forestry, specifically, the federal government's recommendations for making Canada a "model forest nation" (Government of Canada, 1994). Other examples, from medicine and regional planning, will be presented to illustrate the concept of scale. These will include resource allocation, global health, maternal-fetal issues, sterilization of mentally challenged persons, and issues in regional planning, and will constitute the discussion in chapter 4.

· 2 ·

SYSTEMS THEORY AND THE CONCEPT OF SCALE

An Introduction to Systems and Hierarchy Theory

A general consensus has arisen from ecological science supporting a holistic perspective on life on earth (Holling, 1992; Odum, 1989). This perspective is recognized in other disciplines as well (Capra, 1982; Capra et al., 1991; Devall & Sessions, 1985; Faber et al., 1992; Frye, 1982/1987; Gillis, 1990; Haskell et al., 1992; Norton, 1991, 1992; Page, 1992; Potter, 1988; Rolston, 1988; Sagoff, 1992; Sahtouris, 1989; Thomas, 1984). It holds that life and its associations, both animate and inanimate, can be viewed with insights from systems theory as a system of multiple dimensions of systems-within-systems with complex webs of interconnections, interrelations and interdependencies (Capra, 1982; Fischer, 1990; Koestler, 1967, 1978; Lazlo, 1972; Lilienfeld, 1978). The functional processes involved are considered to be dynamic rather than static (Capra, 1982; Capra et al., 1991; Ehrenfeld, 1992; Haskell et al., 1992; Karr, 1992; Norton, 1991, 1992; Sahtouris, 1989). The term *system* comes from the Greek root word "synistanai," meaning "to bring together," and is understood as a set of things that are parts that together form a whole. In the holistic perspective with which I am concerned, neither the part nor the whole is prior to the other, and each implies the existence of the other. While parts may be separate from each other, no part is ever separate from the whole.

Furthermore, it is now recognized that ecological systems are effectively perceived to be hierarchically organized in an open fashion (Capra, 1982; Haskell et al., 1992; Karr, 1992; Norton, 1991, 1992; Schaeffer & Cox, 1992). Each system is seen as unique yet integrated into those systems peripheral to it with interaction flowing in all directions (therefore, open). This perceived structure facilitates dealing with the marked complexity of living systems that are considered self-organizing, self-maintaining, and self-generating units (Callicott, 1992; Capra, 1982; Faber et al., 1992; Hannon, 1992; Nicolis & Prigogine, 1977; Norton, 1992; Page, 1992; Sahtouris, 1989). Within these units, creativity, change, and development are constant features (Bookchin, 1990; Bratton, 1992; Callicott, 1992; Faber et al., 1992; Norton, 1992; Odum, 1989; Potter, 1988; Sahtouris, 1989; Ulanowicz, 1992). There is a correlation between the current emphasis on systems theory in ecology and the view that this knowledge can, and should, be used for both the description and management of natural systems in their entirety (Bratton, 1992; Faber et al., 1992; Haskell et al., 1992; Norton, 1991, 1992; Sahtouris, 1989; Ulanowicz, 1992).

Systems have a number of properties that are important in understanding the nature of hierarchy theory and its application to bioethical issues. For instance, Simon (1962), in his seminal paper "The Architecture of Complexity," defined complex systems as ones "made up of a large number of parts that interact in a non-simple way" (p. 468). In systems of this sort, Simon observed that "the whole is more than the sum of the parts, not in an ultimate, metaphysical sense, but in the important pragmatic sense that, given the properties of the parts and the laws of their interaction, it is not a trivial matter to infer the properties of the whole" (p. 468). Simon suggested that a central structural scheme of complexity is the notion that complex systems are "composed of subsystems that, in turn, have their own subsystems" (p. 468). That is, they are hierarchical in nature because hierarchical organization facilitates the development and stability of complexity in natural systems, whether physical, biological, or social. This dynamic is seen not necessarily as ontological but as epistemological (Allen & Starr, 1982) in that it relates to a natural human tendency to see the world in the context of systems and hierarchies of systems. In other words, "an everyday component of human experience" (O'Neill et al., 1986, p. 38) is the perception of the world we live in as being structured in systems and that these complex systems, in turn, tend to be routinely seen as hierarchical in nature. It is through this hierarchical structure that we are able to describe, understand, and even appreciate the existence of such systems and their component parts. Simon (1962) made the observation that "if there are important systems in the world that are complex without being hierarchic, they may to a considerable extent escape our observation and our understanding" (p. 478). He further observed that the question "whether we are able to understand the world because it is hierarchic, or whether it appears hierarchic because those aspects of it which are not elude our understanding and observation" (p. 478) is a chicken-and-egg type of question. Simon (1962) went on to explain that there is strong reason to accept that evolving complexity does in fact tend to be hierarchic, and O'Neill et al. (1986) reiterated this notion in their work on hierarchy theory and ecosystems. It enables development and promotes stability of complex systems.

Most of us can readily think of hierarchical systems common to our everyday observations of the world around us. We see hierarchic structure in our social systems (individual, family, community, tribe, nation, and so on), and we are all familiar with formal institutional organizations that are clearly structured in hierarchies of parts-within-wholes. For example, businesses, governments, and universities have internal departments generally embedded within broader divisions or faculties, and all within the whole of the organization itself. In the biological sciences, we appreciate the cell, for instance, with distinct subunits or levels of function and structure (nucleus, microsomes, mitochondria, cell membrane, and so forth), and yet also, as only a part of larger structures and functional units (tissues, organs, systems, organisms). In the physical sciences, we identify molecules with subunits (atoms, elementary particles) and also as parts of larger structures (macromolecules, compounds ... earth, satellites, solar-planetary systems, galaxies). In moral philosophy, where concepts rather than physical structures are the central concern, we also can recognize narrower and broader levels of interest, for example, applied ethics, ethics, and meta-ethics.

The exact criteria upon which the hierarchies rest and boundaries are elaborated, however, vary depending upon the nature of the system under consideration. For example, Simon (1962) pointed out that most biophysical systems are described in terms of spatial scale while social hierarchies are often best described "not by observing who lives close to whom but by observing who interacts with whom" (p. 469). These various perspectives on where to draw the boundaries in defining the scales of the hierarchy are reconciled by considering the criteria for boundaries to be temporal in terms of the rate of activity, development, and interaction (Allen & Starr, 1982; Norton, 1994a, 1994b; O'Neill et al., 1986; Simon, 1962). These are the "unfolding" or the "dynamics" of scales and are variously called "process rate," "behaviors," "rate-dependent organization," and, "response time" (O'Neill et al., 1986, pp. 75, 83, 87, 88), or "cycle time" and "natural frequency" (Allen & Starr, 1982, p. 37). These clearly are applicable to biophysical and social phenomena that have physical and temporal characteristics, but they are not as clearly applicable to the levels of thought phenomena found in moral philosophy. Here, the separation of levels seems, perhaps, to be more related to the nature of their concern, such as broad or narrow, long- or short- term.

An interesting and important aspect of biophysical hierarchical systems is that lower scales show more rapid frequency and less stability than higher scales. That is, they unfold over short timeframes. Higher scales, on the other hand, are slow in frequency and are more stable. In other words, their dynamic is more drawn out (Allen & Starr, 1982; Holling, 1992; O'Neill et al., 1986; Simon, 1962). For example, consider the following hierarchical system consisting of leaves, trees, and a forest. In their respective responses to light intensity and atmospheric carbon dioxide, leaf activity is rapid, changing moment-to-moment to accommodate the changes in these parameters. The tree on the other hand integrates the rapid responses of the leaves into growth dynamics that unfold over a considerably longer timeframe that can be read in growth-ring patterns, for instance. The extent and composition of the forest itself, however, is evident only over a period of many years, lasting from decades to centuries. Similarly, in the human social context, we might consider the following hierarchical system: physiological activity, individual, family, community, and society. Here, change and creativity take place moment to moment on the physiological scale, from hours to years on the individual scale, and years to decades on the family and community scale, and decades to several generations on the societal scale.

It is evident that the more slowly responding, higher scales are the context in which the faster responding lower scales unfold. As nested hierarchies, the lower scales are contained within the higher scales, and they "experience" them as background constants to their own activity. Furthermore, the lower scales are constrained by the dynamics of the higher level (Allen & Starr, 1982; Holling, 1992; O'Neill et al., 1986; Simon, 1962). Allen and Starr (1982) elaborated on the hierarchical organization of systems and consider both the context of the system and the content. By context, they mean the connections, interconnections, and interrelationships of the systems' wholes and parts. By content they mean the constraints that characterize these relationships. They declared that "ordered systems are so, not because of what the components do, but rather because of what they are not allowed to do" (p. 11). Physical laws in chemistry and physics are obvious examples that we simply take for granted. Allen and Starr (1982) also characterize the context and content as an information exchange that consists of a signal transmitted by one entity within the system and received and interpreted (that is read as a message) by another entity in the system.

Under normal circumstances, since entities at different scales have different response rate characteristics, there is the tendency for signals from faster rated constituents to be "filtered" (Allen & Starr, 1982; Koestler, 1978) such that the lower scale (faster cycle time) exerts little influence on the higher scale (slower cycle time). However, signals from the slower-rated constituent cannot be similarly filtered by the rapid-rated component, and the message received is therefore taken as a constraint. Koestler (1978) uses the term *holon* to refer to the constituents of a system and indicates that a holon is both a part and a whole. Under normal circumstances, higher-level holons (wholes) generally average a signal from the lower level holon, and so experience a negligible effect on their behavior. On the other hand, the lower level holon integrates the higher level holon's signal over such a short period (spatiotemporally) that the signal is more or less received and acted upon immediately. That is to say, the signal from the higher level is extremely influential on the lower level, while that of the lower level exerts little influence on the higher level. The key concept here, however, is that the higher level scales constrain the lower level scales. While this dynamic clearly applies to biophysical phenomena, it also applies to social phenomena. For example, in the individual / family / community / society hierarchy, where response rates progressively lengthen from individual to society, influence from individual to society is weak while that from society to individual is strong. This concept of scale does not lend itself so well to the hierarchies of thought phenomena, although we may be able to apply it to the notion of enduring thoughts and concepts as distinct from more transient ones. In this regard, the more durable ones may be considered long-term and establish limitations for the more transient short-term concepts.

Allen and Starr (1982) explained that the behavior of the system arises from the "freedom that comes with constraint" (p. 15). That is,

> the constraint gives freedom from an infinite and unmanageable set of choices; regulation gives freedom within the law... the constrained holon is free to do its will within the constrained region... the power to constrain gives the burden of responsibility, whereas being constrained gives freedom from these pressures. The self-assertiveness of the constraining environmental holon in its struggle with its own constraining superenvironment essentially protects the subholons that the constrained environment itself controls. (p. 15)

O'Neill et al. (1986) pointed out that "the most common way to avoid complexity is to over-emphasize a single type of observation set" (p. 7). An "observation set" is considered to be "a particular way of viewing the natural world" (p. 7) and includes not only the particular phenomenon of interest, but also the methods used to measure the phenomenon and the way the measurements will be analyzed. This is clearly paradigmatic in nature (Kuhn, 1970), that is, there is a clear parallel between this concept of observation set and Kuhn's use of paradigm as a model or pattern of beliefs and understandings that establish "coherent traditions" (p. 10) in science. O'Neill et al. (1986) further point out that the space-time scale is a critical component of the observation set, stating that "phenomena must be measured and explained at the relevant spatiotemporal scale" (p. 37). This hierarchical systems perspective suggests that the decision about what scale one will be assessing and considering is an evaluative judgment. Whatever scale is chosen will influence the options available for action and the constraints imposed by the associated higher-level systems in the hierarchy.

> Specifying the proper spatiotemporal scales on which to view the dynamics of the natural world requires ecologists to go beyond their own perceptual levels. They must conceptualize the system on the scale at which phenomena are actually observed. When the proper spatiotemporal scale is designated, the appropriate level of ecological organization can then be made clear. (O'Neill et al., 1986, p. 68)

For example, consider the effects of some perturbation of a natural system, such as fire, excessive precipitation or drought. If, for instance, one is considering the effects of a forest fire, the changes noted within the forest will depend upon the area being looked at. Focusing our attention on one organism within the forest we notice its absence in the area burned and in this area we see destruction and loss of biomass and biodiversity. However, if the whole of the forest is the object of observation, the organism is hardly affected and the area burned may be seen as insignificant. Furthermore, if the timeframe of observation is that of the rate of forest growth, the overall biomass and biodiversity may be seen to be unaffected or even increased by the reintroduction of usable nutrients to the system. It is necessary to explicitly view the ecosystemic response in terms of the temporal characteristics of the perturbation and its effects. The point is that "the system must be defined relative to the scale of the problem being addressed" (O'Neill et al., 1986, p. 85). The context must be made explicit. This is an important notion and, later in the book, I will return to it in terms of the consideration of scale and values.

This clear recognition of the critical importance of the context of the phenomena under consideration is echoed by others (Bratton, 1992; Ehrenfeld, 1992; Haskell et al., 1992; Karr, 1992; Norton, 1991, 1992; Rolston, 1988; Sahtouris, 1989) who stress a contextual approach to deliberations in management and policy matters.

Traditionally, the dominant Western view of the world has been rooted in a positivist, objective-analytic-reductionistic epistemology that recognizes only one acceptable way of thinking, questioning and knowing about the world (Capra, 1982; Kafatos & Nadeau, 1991). This positivist view holds that the world is rationally known only through our senses and that phenomena are objects for study and analysis, best done by reducing them to their parts, through which the mechanics of their functioning are then known. The observer and the object are considered totally separate from the object of investigation, thus providing an objective and quantitative description of reality. Metaphysical concerns, like the nature and source of phenomena, in a qualitative sense, are deemed by the positivist to be "*non*sense, nonexistent" (Sahakian, 1968, 244). Historically, however, there have always been some who have stressed that there are ways of thinking other than reductive-analytic positivism. This post-positivist view encompasses a "shift from objects to relationships" (Capra, 1982) with less emphasis on parts and more emphasis on the relationship of parts and the nature of the whole. By like token then, there is less emphasis on the rational-analytic mode of thinking and more on the intuitive-synthetic mode of thinking. Some recent writers, seeking a more holistic (the properties of the whole are not understandable by reduction to those of the parts) strategy, have integrated concepts of ecology not only into their assessment of ways of knowing, seeking meaning, and valuing (Bookchin, 1990; Capra et al., 1991; Norton, 1991, 1992; Page, 1992; Sagoff, 1992; Sahtouris, 1989), but also into worldviews (Capra, 1982; Capra et al., 1991; Devall & Sessions, 1985; Kafatos & Nadeau, 1990; Milbrath, 1989; Norton, 1991; Sahtouris, 1989), spirituality (Capra, 1982; Capra et al., 1991; Devall & Sessions, 1985; Kafatos & Nadeau, 1990; Sahtouris, 1988), and the conception of parts and wholes, unity and diversity (Bohm, 1980, 1986a, 1986b; Bookchin, 1990; Capra et al., 1991; Kafatos & Nadeau, 1990; Norton, 1992). It is within the varying contexts, interconnections and interdependencies in the natural world that some post-positivist thinkers search for a more meaningful conception of the world around them.

Bookchin (1990), however, pointed out that despite the claim of holism in ecology, it and systems theory generally, are as strongly reductionistic and captivated by a positivist science as are previous models. This reality is evident when one considers the ecologist's effort to characterize relationships in terms of mathematical formula expressing inputs and outputs but still not in tune with the historical, developmental nature of the relationships. However, several writers in environmental policy management have stressed that their approach is a "minimalist" one with respect to their use of systems theory (Haskell et al., 1992). They appear to take, from systems theory, the key idea that a hierarchical organization creates contextual and scalar requirements for policy development, yet still rely on analytical, reductive kinds of systems modeling, particularly with mathematical models. Bookchin (1990) criticizes this, as noted above, and identifies a need for more "speculative thinking" (p. 156, p. 175) that goes beyond the positivist emphasis on concrete facts to allow a dialectic process to occur. He characterizes this "ecological dialectic" (p. 167) as a process of "differentiation rather than conflict" (p. 170). This differentiation "is *development*, not mere 'change'; it is *derivative*, not mere 'motion'; it is *mediation*, not mere process; and it is *cumulative*, not mere continuum" (italics in original) (Bookchin, 1990, p. 170). This sentiment is captured by Sahtouris' (1989) term "mutual consistency." Synthetic thinking (Miller, 1991, p. 33) of this type is a key aspect evident in modern physics and cosmology where a chink in the positivist model initially arose. The view from physics and cosmology now fosters the understanding that uncertainty is a real issue in everyday life. The observer's choice of boundaries for observation determines the context within which observations occur, the questions that are asked and the analyses, syntheses, and decisions that may be subsequently made (Kafatos & Nadeau, 1990). Indeed, the use of hierarchy theory and its emphasis on scale is consistent with these insights about dialectical thinking because attention to scale encourages an explicit identification of parts and wholes, that is, a drawing out of what is implicit and undifferentiated. Reflection on the nature of parts and wholes is informed by the knowledge that the whole is greater than the sum of its parts. The dialogue that ensues in the iterative process of recognizing the nature and values of the different scales seeks the balance that supports the integrity and good functioning of each one.

I suggest that the notion that the whole is greater than the sum of its parts is important for moral deliberation too, and that attention to scale provides some guidance in this regard (see chapter 3). This is so because undertaking a dialogue about an issue where different scales are involved amounts to what I will call a "magnitude error." Since scales are different precisely because they are not the same in magnitude, arguing as if the scales were all equivalent is a mistake. More precisely, the defining characteristics of scales noted earlier, rates of response, development or unfolding, cycling time, and spatiotemporal features, have magnitude differences that enable boundaries to be recognized and, therefore, various scales to be separated and defined. Take, for example, the difference between the development and physiological functioning of a leaf and the development and functioning of a forest. The difference between these scales is many orders of magnitude and is substantive. Comparing and equating them as if there were no difference is unacceptable. Vertical (for the sake of a better term) argument between scales requires the recognition of the magnitude difference and the accompanying implications of this difference, while horizontal argument within a scale does not. As noted earlier, O'Neill et al. (1986) emphasized the fact that a spatiotemporal scale is a critical aspect of any particular way of viewing the world and, therefore, phenomena must be considered and argued at their "relevant spatiotemporal scale." In the context of moral argument, values are such phenomena, and so it is critical to recognize the scales in which various values are most active and have meaning. The choice of values to be emphasized will determine the scale being emphasized and, vice versa, the choice of scale will determine the values to be emphasized. Following hierarchy theory, the contextual scale provides meaning for all its parts and determines which actions are allowable and which are constrained, thus the importance of explicit recognition of scales. When the contextual scale is designated, the appropriate ordering of values becomes clear. Sahtouris (1989) pointed out "that our maturity as a species depends on our accepting the responsibility for our natural heritage of behavioral freedom by working consciously and cooperatively toward our own health along with that of our planet" (p. 24). This working consciously and cooperatively toward our own health along with that of others and of our planet is clearly a scalar endeavor involving parts and wholes. Sahtouris (1989) referred to this endeavor as finding "mutual consistency" (p. 27) or shared harmony with the natural world. This concept is akin to what Bookchin (1990) identified as humanity finding not only our place in nature but also nature's place in us (p. 115) and in his description of an "ecological dialectic" (p. 167), which I outlined earlier.

Hierarchy Theory, the Concept of Scale, and Bioethics

The concept of scale, here extracted from hierarchy theory oriented to systems, has demonstrated its usefulness in dealing with complex ecological systems

(Allen & Starr, 1982; Costanza et al., 1993; Fox, 1992; Haigh, 1987; Holling, 1992; Klijn & Udo de Haes, 1994; Norton, 1990, 1992, 1994a, 1994b; Norton & Ulanowicz, 1992; O'Neill et al., 1986; Simon, 1962; Turner et al., 1995). Since human life is a complex biophysical-social system, there is no reason why this concept of scale should not also be useful in complex human ecological systems. This book supports this concept. The conceptualization of scale presented here, one that incorporates a prescriptive aspect as well as a descriptive aspect (discussed in further detail below and in chapter 3) has not been explicitly adopted in or adapted to bioethics. Perhaps more accurately, the prescriptive aspect of scale proposed here, has not been made explicit earlier. Specifically, Roy et al.'s framework of conflicts in bioethics does not acknowledge scale explicitly as a source of such conflict. There is, however, an understanding of the descriptive aspect of scale in their consideration of bioethics as clinical ethics (individual scale) and as public and social ethics (broader scales). Furthermore, other contemporary texts in bioethics (Beauchamp & Childress, 1994; Beauchamp & Walters, 1994; Levine, 1993; Young, 1989) also appear to attend only to the descriptive aspect of scale and not to the prescriptive aspect described here.

Evidence for scale as a source of conflicts in bioethics is especially noted in the issue of resource allocation in medicine where micro, meso, and macro levels are used descriptively to distinguish between individual, institutional, and societal scales, respectively (further discussed in chapter 4). It is, however, in the environmental field and in ecology where recognition of scalar concerns is most prominent and most strongly brought forward in the development of hierarchy theory (Allen & Starr, 1982; Holling, 1992; O'Neill et al., 1986). Hierarchy theory is a conceptual systems theory with broad application in many fields. The reader is referred to the literature for comprehensive explanations and treatises (Auger, 1992; Pattee, 1973; Simon, 1962; Whyte et al., 1969). In the field of ecology, hierarchy theory has been used to analyze, model, and more fully understand the dynamic relationships in complex natural systems. It is from this domain that I have extracted the concept of scale that I emphasize here. Interpretations for its application in ecological, environmental, and agricultural management policy, upon which I have relied heavily, can be found in the work of Allen and Starr (1982), O'Neill et al. (1986), Norton (1990, 1992, 1994b), Norton & Ulanowicz (1992), Dumanski et al. (1993), and Wolf and Allen (1995).

A critical notion in hierarchy theory and scale, as noted in the previous chapter and which I emphasize here, is that local phenomena unfold more rapidly in time, and are embedded within broader contexts that unfold over longer time periods and greater areas. The broader scale is "perceived" by the smaller scale as a stable, relatively unchanging background, or perhaps is not perceived at all. The broader contextual scale sets parameters within which the more rapidly unfolding scales experience freedom of action. Thus, when a trend in activity at the local, less stable scale becomes great enough to overwhelm the ability of the broader, more stable and slower compensating scale to compensate or adjust, destabilization of the whole system is expected. Furthermore, change and development in the global scale are expected to be accompanied by change in the local scale since the latter is dependent on the former (Allen & Starr, 1982; Holling, 1992; Norton, 1990, 1992, 1994b; Norton & Ulanowicz, 1992; O'Neill et al., 1986).

An understanding of scale derived from hierarchy theory (Costanza et al., 1993; Fox, 1992; Haigh, 1987; Klijn & Udo de Haes, 1994; Norton, 1990, 1994a; Norton & Ulanowicz, 1992; Turner et al., 1995) aids in interpreting the dynamics ingredient in Norton's observation that "human values and choices are becoming more and more dominant drivers of ecological change" (Norton, 1994a, p. 12). Norton further observed that these same human values and choices drive all of our social and scientific policies. Each scale, whether inter-human or extended beyond to the man-in-nature realm, has its own predominant values and these values need to be sorted out and explicitly acknowledged. For example, following from Norton's observation quoted above, we can consider values that reflect concern for multiple generations of humanity. This concept includes the recognition that "we must act as members of natural as well as the human social community" paying attention "to the context in which our values are formulated and acted upon" (Norton, 1994a, p. 16).

Hierarchy theory suggests that each of the shorter time-frame and more localized scales operates within the context of a longer and larger scale, and therefore, experiences certain constraints dictated by the broader scale. Freedom within any scale is contingent upon the constraints of the broader contextual scale (Allen & Starr, 1982; Dailey & Norton, 1994; O'Neill et al., 1986). Explicit attention to scale implies that all scales should be considered. Knowledge about each scale, including which scales are contextual and which constituent (that is, whole and part) can and should be used to inform any dialogue regarding actions. Where knowledge is lacking, it is sought and asking questions about scales helps guide the knowledge that is needed. This is the descriptive aspect of scale more fully discussed in chapter 3. When uncertainty remains, emphasis on the longer and larger scales that give context to the others is favored. This is the prescriptive aspect of scale.

It is necessary at this juncture to elaborate a little more on my understanding and use of the concept of scale to be certain that it is clear. There is a tendency to equate "scale" with "level," but I interpret the two to be quite different. The use of scale in ecology has a distinct, multi-dimensional character, taking into consideration the three dimensions of space and the dimension of time. In fact, as explained earlier, the division into different scales is most closely associated with temporal rates of activity, or what I have referred to as the unfolding of the life of the scale under consideration. This understanding of scale is most consistent with *Webster's Dictionary* definition of scale as "relative size," a ratio of size and "a series of graduations" and *Oxford Dictionary of Current English's* (2006) definition as "a range of values forming a system for measuring or grading something." The concept of level, however, seems to me to be distinctly two-dimensional, not at all what I am suggesting with regard to scale. This is supported by *Webster's Dictionary*, where level is defined as having the characteristic of being "horizontal" and the *Oxford Dictionary* definition of level as "a horizontal surface" or "a position" or "rank." All of the synonyms offered by *Roget's Thesaurus* (1962) also imply a two-dimensional view for the notion of level that does not capture the depth and complexity of the structure of scale as I intend it to be understood. For example, let us consider the scale of a human individual. This scale has a distinct physical boundary defined by the surface of the body, perhaps two square meters. In temporal terms the activities of this scale take place through time-frames of milliseconds (for its component electrochemical activities) to days and weeks (for its organ systems) and up to several decades (for its experience of a life-time). The rich multi-dimensional nature of this scale is obvious, and the complexity of the relationships and interdependencies of all of the parts or subsystems that make up the whole of this scale is evident upon consideration. This individual scale, however, does not exist in isolation but experiences its life in a larger context. That is, in descriptive terms, it is a part, a subsystem, of some larger whole within which it is in relationship with other parts. These relationships, and their spatial and temporal characteristics, define for the individual a broader contextual scale upon which it is dependent and within which it is unfolding, living out its life. Referring to these respective scales as levels does not adequately capture the multi-dimensional complexity that is implicit in the nature of scale.

Again, as noted above, the concept of scale discussed here applies to biophysical and social phenomena where spatial and temporal characteristics are more readily identified. This concept of scale adopts response time/rate of development as the means of establishing boundaries between scales. I have not intended to apply this concept of scale to purely mental constructs that are less amenable to such a criterion for establishing scalar boundaries.

Current Views of Scale

Ecological Scales

Norton, in a series of workshops (1994a, 1994b), internal discussion papers (1994a, 1994b, Daily & Norton, 1994), and in published material (1990, 1991, 1992; Norton & Ulanowizc, 1992) has suggested the application of hierarchy theory to the environmental policy-making process. He expounded the usefulness of hierarchy theory in describing the nature of natural systems (Norton, 1990, 1991, 1992, Norton & Ulanowizc, 1992) and also prescribing human interaction within them (Norton, 1992, 1994a, 1994b). Norton proposed a scalar nature to human values and defined three basic scales useful for policy deliberation, each scale corresponding to temporally distinct policy horizons: individual, community, and global. Although scale definition is somewhat arbitrary, this is in keeping with Fox's comment that "scientists from different disciplines recommend the use of three scales, for example large, intermediate, and small" (Fox, 1992, p. 293). Norton's individual scale corresponds to a micro scale defining spatial relationships in terms of centimeters to tens of meters and temporal relationships in terms of days and decades. This local scale is dominated by activities that unfold at rapid rates over short time periods and local space characteristic of living processes that proceed through short life cycles. In human terms, this local, individual, micro scale is the one in which individual people live out their lives and make their life choices within established constraints and rules, be they natural, legal, economic, or moral. The middle or meso scale is characterized by activities unfolding in terms of tens of meters to hundreds of kilometers in the spatial dimension and years to many decades in the dimension of time. This intermediate scale incorporates natural life processes occurring at scales commonly referred to ecologically as the community, or the landscape, or the bioregion. In terms of human activity, Norton describes this as "a longer and larger community oriented scale in which we hope to protect and contribute to our community which might be taken to include the entire ecological community, ... one in which we feel concern for our cultural connection to the past and the future" (Norton, 1994a, p. 15). Norton further characterized this scale as a multi-generational one in which "we protect, develop and nurture our sense of who we are as a culture ... and decide what kind of society we want to be" (Norton, 1994a, p. 15). These considerations clearly have normative implications. Norton continued describing this scale as the one where,

> we feel concern about the culture's interaction with ecological communities that form its context. The second scale is doubly important because it corresponds roughly with the ecological time scale in which multiple generations of human individuals, organized into communities, must relate to the populations of other species that share our habitat. (Norton, 1994a, p. 16)

As Norton (1994a) well expressed, it is at this scale that human activity has its greatest impact on ecologic systems and where intergenerational considerations are critically important and require "special attention to large scale aspects of the landscape" (p. 16).

A macro scale defined in terms of hundreds to thousands of kilometers in space and centuries to millennia in time is defined by the slow-rate processes of geology and evolution. These are essentially so slow that they are seen as constants from the perspective of the activities unfolding in the micro and meso levels. This global scale from a human perspective is one in which "humans express hope that their own species even beyond current cultures will survive and thrive" (Norton, 1994a, p. 15).

Using this formulation of scales as a framework, Norton and Ulanowicz (1992) have shown that protection of biodiversity is best modeled at a landscape ecosystem level, which is a meso scale, space-time dynamic. Policies that do not focus on this scale, but instead focus on the micro scale of individual and corporate preferences, economic and free market values, will be inappropriate (Norton & Ulanowitz, 1992). Also, using this model of scale, Daily and Norton (1994) defined three scales of public values defined as individual, intergenerational, and evolutionary, corresponding to the micro, meso, and macro scales. They suggest that management objectives for forestry policy can be defined by ascertaining the values critical at each of these scales and by understanding that the larger and longer scales (that is the macro and meso) place action constraints on the shorter more local level scales.

Societal Scales

It has become the routine in medicine to consider resource allocation on three levels; microallocation, mesoallocation, and macroallocation (Canadian Medical

Association, 1994; Hiller, 1984; Lemieux-Charles & Meslin, 1993; Manga, 1987). This view of resource allocation issues in medicine clearly reflects a scalar approach. Hiller (1984) proposed three levels (or scales in the terminology of this study) of concern when undertaking analysis of various organizations. Following the lead of the resource allocation debates, Hiller identified the three levels as micro, meso, and macro. The micro-level is that involving the individual and his/her values; the meso-level considers the institution and its values; and the macro-level takes in the broader community and societal values. This organizational framework has now been widely adopted (Canadian Medical Association, 1994; Lemieux-Charles & Meslin, 1993) in various areas of analysis in medicine and policy development and, as noted, particularly in the resource allocation issue.

It is clear that this framework provides some conceptual clarity to the debates, but it is not understood in the context of hierarchy theory and scale as I have elaborated them. Basson (1979), for instance, states:

> Medical resource allocation has traditionally been divided into microallocation—selecting among individual candidates ... and macroallocation—determining on a societal level how our health care dollar should be spent. A morally significant distinction between the two has yet to be demonstrated. Microallocation is just macroallocation on a small scale. (p. 313)

He goes on to consider them interchangeable. From the perspective of hierarchy theory, however, they are not at all the same. Ecologically, hierarchy theory and the concept of scale that is integral to it consider the micro scale to be embedded in the macro scale. More specifically, in ecology, the macro scale is the contextual scale for the micro scale, and the facts (e.g., availability of a resource) of the macro scale set constraints on the micro scale. Although human sociocultural systems are not entirely like ecological systems (Westra, 1994), the concept of scale and its concern for context, as shown above, are quite applicable to sociocultural systems, and the various scales that can be appreciated in a sociocultural system are not interchangeable concepts. Scale uses rate of development/response time as criteria for boundary determination and socioculture systems have these characteristics. Therefore, such systems are amenable to scalar analysis. The facts and values of all scales involved are valid concerns. Furthermore, the facts and values of the identified contextual scale can be seen as constraints on micro scales and, therefore, are morally significant. In human sociocultural systems where values are driving forces, a scalar view of values, that is, ascertaining the values embedded in the various scales, carries with it the implication that the values can not be simply argued or compared without being clear about the scale within which the specific values arise. Such an approach enables the recognition of values that are shared by the different scales as well as those that differ from scale to scale. The scalar approach, by definition however, implies that the values of the whole (higher scale) subsume those of its parts (lower scales). For example, the intermediate scale described earlier includes the economic values of its constituent local scale but has, as well, broader value concerns, like natural diversity, which are not concerns of the local scale based on boundary definitions arising from response rates as outlined above. Not identifying the scale creates a cross-scalar (vertical coupling) debate and the previously noted magnitude error. From a scalar perspective, higher scales set constraints under which lower scales must develop and optimize their function. Therefore, in a cross-scalar values debate, errors in magnitude, that is the nature of the scales is not recognized, could lead to the failure to recognize constraints that may be critical to the integrity, good functioning, and even the very survival of the lower level scales. Regarding errors in magnitude, scales and the value sets integral to them are different precisely because they are not the same in magnitude, and, therefore, arguing as if the scales were all equivalent would be a mistake. The example above serves to demonstrate this problem. A value like diversity is active in a scale that is higher than the scale in which economics is active and thus diversity sets limitations on the economic activity in the lower scale. A practical example of this dynamic is the desire to constrain the logging of the tropical rainforests, especially in Brazil. Here, the free running economic value of merchantable timber and profit will eventually undermine itself and fail when the rainforests are depleted. Clearly, taking into account the larger contextual scale of the rainforest ecosystem and recognizing its diversity, integrity, and good function as limiting values on the economic value will reduce profit over the short term. However, such constraining action will optimize profit over the long term, since it will enable sustainable logging and thus profit, so is really a form of heightened self-interest. The scalar approach validates the intra-scalar (horizontal coupling) debate and requires identification of contextual and constituent scales (wholes and parts) in order to recognize when a cross-scalar values debate is occurring.

Hiller's framework of scales for analysis is a useful one and, if considered from the perspective of hierarchy theory, will provide a new insight into how such scales may be used to distinguish between and among values at the various scalar levels. Making these distinctions helps greatly in clarifying the ethical aspects of decisions that need to be made.

Explicit attention to scale requires that the moral agent recognize and describe clearly the various scales involved, and the values integral to each of them. It requires the moral agent to seek conceptual clarification of what constitutes the contextual and constituent scales, in other words, wholes and parts. The prescriptive aspect of scale (see below) then provides guidance to the moral agent in that contextual scales and values are generally given precedence even at a present cost, while constituent scales optimize their integrity and good functioning within the constraining value-sets of the contextual scale. The impact such decisions have on people, their lives and relationships, and on our society and its functioning, go beyond simple practicalities and economics and take on moral importance.

Hiller's (1984) framework provides a scalar view that is limited to the human social condition—micro (individual), meso (institutional), and macro (societal-governmental). This framework needs to be integrated in some way with Norton's framework to provide a broader basis for analysis and synthesis in bioethics that is global in nature and takes into account the world outside human society. Recognizing scale as a dimension of issues in bioethics provides another form of guidance in choosing between values in conflict. It does so because it carries with it the need to emphasize and generally give precedence to the values/principles of the contextual scale as constraints or limitations on the activity of constituent scales. It would be of enormous help to have a framework for explicitly considering scales. Such a framework would allow the moral agent to determine descriptively what level any particular activity or issue falls into and further, determine the contextual level(s) in which it is embedded and whose values therefore set constraints that provide normative parameters. This is particularly evident in the realm of ecology as noted in the previous section of this chapter and is clearly evident in the discussion above.

Another example of scale in the social context is that of community resource management (Fox, 1992). In his paper "The Problem of Scale in Community Resource Management," Fox makes the following observation:

> Community resource management programs operate on the premise that resources are managed best when the people affected by decisions participate in the design and implementation of these decisions, but finding common ground between government managers and local users of public-domain resources is difficult. Governments seek to improve the welfare of the district or nation, while villagers seek to survive as a community. Planners need data that have been aggregated by administrative areas (counties, provinces, planning regions), whereas villagers are concerned with the performance of households and the use of individual pieces of land. Bureaucrats feel pressure to expand quickly from localized pilot projects to broader regions and have difficulty in dealing

> with local idiosyncrasies, while villagers are concerned only with the local project and their own idiosyncrasies. To be sensitive to the various spatial perspectives from which nations and villages view their resource management problems, planners need to operate on different spatial (and sometimes temporal) scales and to exchange information among these levels. (p. 289)

Regarding the nature of information, Fox notes,

> because information is scale-specific and data collected at one scale should not be used to make conclusions about phenomena occurring at different scales, it is usually necessary to use multiple scales to describe any environmental process completely. ... While determining the appropriate scale classes forms a major subject of investigation, scientists from different disciplines recommend the use of three scales, for example large, intermediate, and small. (p. 293)

Examples where such scalar concerns are evident are numerous in New Brunswick where, in recent years, we have experienced extensive public debate and disagreement over clear-cutting as a forestry practice and, specifically, the proposal by private logging companies to clear cut the Christmas Mountains, the last remaining old growth forest in the Province. Here, the environmentalist community seeking ecosystem "integrity" (Westra, 1994) is pitched against the logging industry seeking increased capital, the loggers seeking employment and security for their families, and the government, theoretically, looking to support, or at least moderate between, all these. The scalar view of this situation requires a clear elaboration of the scales of concern for each phenomenon (individual foresters, logging industry, government, and the Christmas Mountains old growth forest). Following this interpretation, scalar analysis requires explicit recognition of what constitutes the critical contextual scale(s) and the values that are integral to their continuance and fulfillment of optimum development and function. Taking the above case as an example, an individual forester with access to the Christmas Mountains, a forest commons, is concerned with logging this forest, a local scale activity, for immediate economic gain, a local scale value. The logging companies with access to the same forest will have the same concerns and values as the individual forester, and the two may even be competing to maximize their economic gain, thus putting more pressure on the forest ecosystems. Government, the holder of the commons, is concerned with protecting and promoting economic gain, a local scale value, as well as maintaining the commons. Maintaining the commons, the Christmas Mountains old growth forest, is, in scalar analysis, an intermediate scale activity and its integrity and good functioning, including things like biodiversity, are values critical to this intermediate scale. By scalar analysis, these value concerns supersede those of the local scale, assuming that local scale activity is such as to dangerously threaten them. In this way the Government, controller of the commons, has reason to promote the integrity and good functioning of the Christmas Mountains old growth forest as a limitation on the logging activities of individual and corporate forestry enterprises. This means, of course, that the economic value sought by these foresters is to be optimized within the limitations as opposed to being maximized. Therefore, this is not seen as a rough riding over personal or corporate rights and self-interest, but rather, an enabling of a more expansive view of what constitutes their self-interest.

Correlative Ethics

The concept of wholes and parts as a framework for scalar analysis is a definite possibility. Such a framework requires the explicit identification of the parts and wholes in any issue, each of which carries its own set of facts and values and horizon of concern. This approach is similar to Callicott's (1979, 1986) description of ethics as "correlative." He stated, "ethical relations and social organizations are correlative in two ways: (1) the perceived boundaries of a society are also the perceived boundaries of its moral community and (2) a society's structure or organization is reflected in its ethical code of conduct" (1986, p. 406). In this view, as the perceived boundary of human activity and interest expanded from the individual and family unit to the tribe, community, nation, and so on over evolutionary time, so too the boundary of ethical considerability expanded. In other words, the extent of who and what count as morally significant in our ethical systems also expanded. The concept of scale, explicit consideration of parts and wholes and their inter-relationships, is similarly correlative and contextual in nature. Within human social life, for example, the various social organizations and boundaries that contain them are the context in which and from which meaning is elucidated for the contained scales. When the parts and whole framework is extended beyond the strictly human field to include the broader natural world, an integration of ecological and social scales occurs. This correlative view of ethics can be represented as nested concentric circles (or spheres if you wish to give it depth) (Glaser, 1995).

Proposed Framework of Scales

In the section above, I reviewed several current views of scale. Briefly, these were Norton's individual, community, and global scales addressed from an ecological perspective, Hiller's micro, meso, and macro scales which I have categorized as stemming from a social perspective and Collicott's correlative scales arising from humanity's expanding circles of moral inclusiveness. There is a need to integrate these in some conceptually useful way because their common focus is the philosophical concern about the relationship between parts and wholes. In this book, parts are considered constituents and wholes are considered contexts. Since the terminology for different scales is variable, selecting one set of terms from the literature would be confusing and misleading. Therefore, I propose a new set of terms for three specific scales and a way for considering them from the perspective of constituents and contexts. I propose three general scales formulated from the ecological concept of boundaries determined in a general way by the respective rate of development or response time as described earlier in this chapter. I propose to consider a local scale, a communal scale and a global scale. The local scale is the short-term, here and now, individual scale in which the response time/rate of development is rapid, measured from seconds to a few decades. The communal scale is the intermediate-term, community/society scale in which the response time/rate of development is intermediate, measured in several decades to many hundreds of years. The global scale is the long-term, geo-evolutionary scale in which the response time/rate of development is slow, measured in thousands of years.

Integrating these three specific scales with the concept of context and constituent aspects for each scale gives the following terminology that is easily depicted in this matrix.

Matrix of Scales

	CONSTITUENT	CONTEXT
GLOBAL	Global-constituent	Global-contextual
COMMUNAL	Communal-constituent	Communal-contextual
LOCAL	Local-constituent	Local-contextual

Now, in using this matrix for a scalar analysis one considers any particular phenomenon of concern to be a part of some greater whole, that is, as a constituent, and proceeds to determine whether the constituent's scale of existence is local, communal or global. This assessment is based on the predominant rate characteristics of the constituent under consideration. For example, cutting down all the trees in a forest is clearly different from cutting all the leaves off a single tree or all the leaves off all the trees in a forest.

The response rate, or in this case the rate of development, of each of these is different. The development of a leaf occurs rapidly over a few weeks in the spring and then remains actively functioning for the season, that is, several months, and then the leaf dies and falls away. The development of a tree is a much slower process taking many decades to reach maturity, but then continues many more in a state of maturity. The development of a forest is even slower, taking several decades or even centuries to establish itself and then may remain in a stable state for many more. Therefore, the leaf, or leaves, would be considered local in scale; the tree is also local in scale while the forest is communal in scale. The leaf and tree are clearly different in their respective rates of development and it is clear that the tree would be considered the context for the leaf, that is the broader scale that imparts meaning to the constituent. Thus, the tree would be considered the local-contextual scale and the leaf the local-constituent scale. Similarly, the forest, by definition as a communal scale object, would be considered contextual for local scale objects of concern. This is entirely consistent with our view of forests as relatively stable, long-lasting entities that are made up of trees and leaves.

As a simple and obvious example, imagine a small lake on which there are four people living in homes and disposing of their human waste directly into the lake. Now the lake is big enough to accommodate this amount of human waste, buffer its effects and, from the standpoint of the inhabitants and the ecologic community, not suffer in any way. Using our matrix, the human activity is unfolding in a local, individual scale (local-constituent), while the lake and its ecology are unfolding at a broader ecosystemic scale (communal) and are not unduly affected by the activities of the four individuals. However, if there were to be an influx of individuals (local-constituent scale) living on the lake all dumping their waste into the lake, then there will come a point when the natural systems (contextual scale) will not be able to accommodate this amount of waste. Now the lake will become polluted, and the natural systems (contextual scale) will suffer and the individuals (local-constituent scales) living on the lake will suffer too.

Here we see the facts of the natural capabilities of the lake ecosystems (communal scale, contextual to local scales) setting constraints (natural ones in this instance) on the individuals living on the lake (local-constituent) with regard to how many can do so while still dumping waste into the lake before the lake conditions (the context) change unacceptably. The local scale activity is constrained by the communal scale, if the value of clean unpolluted lake water and an associated healthy ecosystem (contextual scale, which includes the people living on the lake) is to be supported. Individuals, if they want the lake to remain pure, must recognize and give moral weight to these constraining values. They might for instance band together as a community organization (local-contextual scale) and set clear policy (constraints) on the acceptable number of inhabitants or that all must change their dumping of waste habits. This is analogous to the case described earlier involving the Christmas Mountains. Recognizing that there are different scales of concern here, and that each has its own specific necessities, would help in considering the actions individuals should take to support the identified necessities.

In terms of medicine and clinical ethics (that deal with questions of what should be done regarding a specific patient, an individual), the object of concern is the individual. This concern would be considered a local-constituent scale of activity, while the broader horizons of concern like family, extended family, and public community would be considered to be local-contextual scales. Social systems and nations exist over periods of time consistent with classification as communal scales and are the context in which the local scale exists. These broader objects of concern would be considered communal-contextual scales.

It is my contention that this framework of scales enables a conceptual approach to addressing scale in bioethical issues. I propose the following questions as a structured framework to be used to approach scale and enable the concept of scale to enter deliberations in bioethics.

> What scale, global, communal or local, is the phenomenon (action, object or value) of concern occurring in?
> What is (are) the contextual scale(s) and what the constituent scale(s)?
> Is the integrity of the contextual scale at risk?

With this framework of scales and questions in mind, it is easy to conceptualize dialogue and argument as horizontal, occurring within a scale, and vertical, occurring between scales. This approach to scale requires the moral agent to define the scales involved based on their characteristics, that is, local, communal, global, context, and constituent.

To summarize the points made above, the concept of scale has not been explicitly recognized as an issue in bioethics. A great deal of effort is expended on "values dialogue" without fully appreciating that a choice of values priority is also a choice of scale priority which may place us in a scale which is inappropriate for the phenomenon (issue) being considered. Roy et al.'s (1994) framework for considering conflicts in bioethics does not explicitly acknowledge scale as a source of conflict. However, as noted above, there is an understanding of this as a descriptive factor in considering bioethics as clinical ethics (local-constituent scale) and as public ethics (local-contextual scale), and social ethics (communal scale). Expanding beyond hierarchy theorists' (Allen & Starr, 1982; Holling, 1992; O'Neill et al., 1986; Pattee, 1973) use of hierarchy theory in analyzing, modeling and understanding dynamic relationships in complex natural systems, Norton and others (Dailey & Norton, 1994; Giampietro, 1994; Norton, 1991, 1992, 1994a, 1994b; Norton & Ulanowcz, 1992) have recommended its use in modeling to aid policy making. I believe that the concept of scale from hierarchy theory can be integrated into bioethical deliberation and used in analyzing the complexity of human relationships, activities and values. It provides not only a descriptive approach to bioethical conflicts but also a prescriptive approach (both further elaborated in the next chapter). The prescriptive nature of the concept of scale directs further inquiry and gives guidance to a values dialogue with emphasis on understanding the importance of contextual relationships in which the scale of concern, its contextual whole and constituent parts, need to be explicitly identified.

· 3 ·

THE MORAL NATURE OF THE CONCEPT OF SCALE

Description: Wholes and Parts

It is necessary for us to recognize conceptually the notion of scale for it to be seen as an issue in bioethics. I would now like to address the implications of the addition of this concept of scale to our framework of issues in bioethics. Emberley (1995) described Aristotle's account of ethics as political, teleological and cosmological (p. 17). For the purposes of this work on scale in bioethics, I am focusing on the cosmological nature of ethics. By cosmological, Emberley (1995) means that:

> The particulars of ethics correspond to both the structure of existence itself and the human effort to actualize that structure as the permanent foundation of human order. Thus, ethics is not simply a functional requirement of social life, or an instrument for more effective satisfaction of wants. There is a cosmological dimension to ethics; acting ethically recalls the structure of existence itself, and actualizes and perfects the human experience of it. (p. 21)

I suggest here that attention to scale, as elaborated in the forgoing chapter, does precisely this. I suggest that attention to scale and cosmological dimension in ethics is much like creating a narrative, particularly "stories of identity" (Gardner, 1995/1998) that incorporate meaning, "convey values, build esprit de corps, create role models, and reveal how things work ..." (Stewart, 1998). Epistemologically, explicit attention to scale is like developing a story about interdependency and interrelationships, thereby revealing the structure of our human existence in interpersonal, social and geo-ecological terms. This view is closely linked to the conceptions of reality now understood by physicists. Bohm (1980, 1986a, 1986b), for instance, is concerned with the singular wholeness of the universe and sees this "enfolded" into the parts or subwholes in such a way that it can be perceived by an "unfolding" of these units. He refers to this conception of reality as the "implicate order". Explicit attention to scale also stresses the age-old philosophical and ethical concern with wholeness, wholes and parts. The emphasis on the concept of scale provides both a description of the complex relational organizations of reality and, in conjunction with the demand to place human activity within that descriptive framework, a prescription or guide for human behavior. This I will address in the next section.

The jump from description to prescription, which is proposed here, may seem to fall into what has come to be known as the naturalistic fallacy. That is, the perceived error of deriving an "ought" from an "is". It is generally understood, in philosophical argument, that it is logically incorrect, and therefore a "fallacy", to expect to be able to derive a "statement of value" (that something is of great worth, is highly cherished and respected and therefore ought to be protected and promoted) from a simple "statement of fact" about the way the world is. However, as Norton (1994a) observed, "the recognition that some of our concepts are morally thick is really only a special case of a much more general phenomenon—description and prescription are so entwined in our use of language that it is almost always impossible to separate them in ordinary discourse" (p. 4). He is suggesting that it is not a simple matter to say that facts and values are two different kinds of premises or concepts. In reality, they are so integrated that it is often impossible to clearly separate them, so moving from one to the other in the course of consideration is unavoidable. For example, consider the following words: brave, honest, healthy, sick, beautiful, ugly. They all have both a descriptive (factual) character, as in describing someone as honest, and a prescriptive (ought to) character, as in acting honestly is a preferred mode of behavior. Similarly, consider these words: stability, diversity, balance, integrity. They are descriptive of ecosystem states, and from the perspective of one looking at the functioning of an ecosystem, they are also desirable states for the ecosystem to experience. As Rolston (1988) said, "an ethic needs a grasp of reality. The way the world *is* has a bearing on the way it *ought to be*" [original emphasis] (p. 74). Rolston further commented that, in the context of ecosystems and scales, "crossing over from science to ethics, we can discover ... the values in such a community-system and ... our duties toward it. Interdependence does not always deliver duty, but biological obligation is a relevant consideration in determining moral obligation." (p. 173). In other words, in the realm of ecology (and all things are part of an ecology), what "is" establishes the reality, and therefore, the constraints or limitations on what "ought to be."—This is echoed by Westra (1994) in defense of her "principle of integrity." She stated, "... it is sufficient that the 'is' should be seen as a limit to whatever 'ought' we might be able to defend" (p. 99). This doesn't mean that "ought to" statements come directly from "is" statements, but that they are integral to each other and the "is" statement is a priori.

I understand a value to be a form of belief about what things/ ideas/ needs/ state of affairs one considers important, prized, cherished, preferable or desirable (Carbone, 1987; Marietta, 1979; Reich, 1995; Sankar, 1992). It appears to me that values are actually emergent properties not deduced or derived from facts ("is" statements) but instead emerge (unfold using Bohm's term) from them. The values are enfolded in the facts and become evident as these are unfolded (experienced and observed), and the ought statements extending from them become clear as we consider the values in the context in which they have meaning and in which we also find ourselves. This evaluative process takes place in the human mind and heart, that is, it is both cognitive and emotive, and is as yet a mysterious black box. This view is supported, as Marietta (1979) points out, by the phenomenology of perception where it is recognized that one's moral judgment is more than a rational deductive exercise. In this field, referred to as the intentionality of consciousness, "the active role of consciousness in attending to objects and events and in constituting meaning in them as they are perceived in the world" (p. 200) is taken into account. From this perspective, consciousness is the black box, the central constitutive phenomenon, giving life to one's experiential observations of both facts and values simultaneously. Rolston (1988) suggested the same idea, namely, "for some, at least, the sharp *is/ought* dichotomy is gone; the values seem to be there as soon as the facts are fully in and both alike seem properties of the system" (p. 232) (original emphasis). Of course, we know that the facts are never really "fully in," but we do function from the premise that we want all the data that we can reasonably have. Marietta (1979) further points out that this fusion of fact and value cognition in our experiential observations of the world is, only later, broken by critical reflection. He characterizes this subsequent reflective separation of fact and value as a "second level interpretation of our seeing" (Marietta, 1979, p. 200) since it is not a constituent of our immediate experience of the world. This evaluative process, the move from "is" to "ought to," is dynamic and in constant operation. It is an "internal model" (Marietta, 1979) where the factual and value descriptions lead naturally by reflection to the embedded ought that is consistent with the world as observed and experienced. Decision-making is possible only by this dynamic evaluative process through which some aspect of what is becomes a part of the "web of human meaning" (Emberley, 1995, p. 17), thereby taking on some value for us from which action occurs.

The exercise of explicit consideration of scale and sorting out the concerns of wholeness, wholes and parts, is to be seen as part of this dynamic operation of discovering the emergent ought from the existing is. This attention to scale is a component of the evaluative process that creates the web of human meaning mentioned above. Different conceptions of ought are not clearly deduced but rather emerge as the lived experience unfolds.

Beauchamp and Childress (1994), in the fourth edition of their text *Principles of Biomedical Ethics*, emphasize a coherence form of moral analysis and justification rather than a deductive or inductive approach. However, the introduction and use of ethical principles to guide decision-making often results in a deductive form of analysis. In such a deductive approach, what Beauchamp and Childress refer to as "the covering-precept model" (p. 14), there is an existing framework of moral precepts, principles, and rules that are applied to a moral decision in a "top-down" manner and the proper decision logically follows from the precepts.

In an inductive approach, what Beauchamp and Childress (1994) called "the individual-case model" (p. 17), moral reasoning and justification start with the context, circumstances, and particulars of an individual case and proceed from the "bottom-up" to elaborate the rules and principles that seem to be operational in this instance. In the coherence approach, analysis does not proceed exclusively from either bottom-up or top-down but very deliberately "in both directions" (p. 20). In practice, it is necessary to encourage participants in the bioethics education sessions that I facilitate to take this approach rather than focusing solely on principles and their priority. Perhaps this concept is related to the ease with which we can fall into a mathematical interpretation of the idea of "balancing" (Young, 1989, p. 37) competing claims. Alternatively, perhaps, it is related to the greater comfort people have in talking about the nature and relative weight of a principle than the decidedly uncomfortable feelings generated when talking about the sensitive and complex emotions, beliefs, values, and relationships of a particular individual's case. However, it is in the lived experience of an individual, the depth of their interrelationships, beliefs, and values that we find the core resource for the resolution of the issue, whatever it may be. Attention to scale encourages this coherence approach by looking to identify the whole and the parts, and defining the story of their interrelationships. That is, asking the question about scale explicitly brings attention to and conversation about wholes and parts, the relationships that are involved and what they mean in the context of the circumstances under consideration. Here, the interest in meaning encompasses the embedded beliefs and values and the scale in which they are expressed.

In a narrative sense, this is much like looking at the issue as a story with a plot and numerous subplots (Randall, 1995). Attention to scale, as narrative, encourages us to seek a deeper understanding and appreciation of the ways in which the characters interact, are dependent upon and responsive to each other. Through the narrative's plot and character development, actions, reactions, motivations, intentions, means, and consequences become evident. The elements of motivation, intention, consequences, and means are important determinants of action (Young, 1989) and are of moral significance. Such "narrative ways of knowing" (Charon et al., 1998, p. 558) require the reader to exercise literary skills, "to respect language, to adopt alien points of view, to integrate isolated phenomena (be they physical findings or metaphors), so that they suggest meaning, to organize events into a narrative that leads toward their conclusion, and to understand one story in the context of other stories ..." (Charon et al., 1998, p. 557). These authors, elaborating on the narrative method, explained that it focuses "attention on the storyteller's attempt to find causal or meaningful connections among events, on temporal orderings and reorderings of events" (p. 558) and that storytelling enables "pattern-finding and meaning-making" (p. 558). They further commented that, while the analytic approach to ethics enables a tendency to reduce moral conflicts to "rational problems to be solved" (p. 559), a narrative approach enables a tendency to draw out one's "moral imagination" leading one "to contextualize and particularize ethical issues" (p. 559). They concluded that "attunement to the right and the good is attained by imaginatively rendering, for oneself, the situation of others" and that a story approach "challenges the reader to measure up to another's mode of comprehending the world" (pp. 559–560). Clearly, both of these approaches have merit and a balanced use of each is appropriate. Attention to scale draws us into this narrative way of knowing by requiring explicit attention to description of wholes and parts, to perspective, and to context and therefore helps by enabling a balanced approach to the moral dialogue. For the purpose of description, consider the following stories:

> George Smith has just returned to the kitchen of his home after doing some gardening which he finds a most pleasurable activity and which enriches his life. Today, however, he has been able to spend only a short time in his garden because of the severe pain he has in his right knee and the relative instability he experiences when walking on uneven terrain or when he tries to squat. Although he is only 58 years old, he has been told by his doctor that he has advanced osteoarthritis in his right knee and his left is not too far behind. George finds that he now needs to restrict his activities quite markedly to accommodate his knee pains and has noticed that this has resulted in a general decrease in his energy levels as well as a definite weight gain which has in fact made his knees more symptomatic. George is worried that he will not be able to continue working if he doesn't get something done about his knees, and he is waiting to see an orthopedic surgeon. He hopes this will happen soon since he'd like to get on with the knee replacement surgery which his own doctor says he needs.

> Bob Green sighs a big one and returns his attention to the knee he is operating on. Bob finds his orthopedic surgery practice to be consuming him. He has a long waiting list, not only for patients to see him in his office, but for them to get their subsequent surgery as well. The list is so long that he worries that his patients are suffering and this takes a big toll on him. Today, when he finishes this surgery, he will be participating in a meeting in which he knows he and his orthopedic colleagues will be told that they are very close to using up their budget for total knee implants. Bob worries about what this will mean for his patients, not only those on the waiting list for surgery, but also those waiting to see him in the office who will be added to the waiting list. What can he tell them? What should he tell them? Bob and his colleagues are trying to get their prosthesis budget increased, and are looking at other ways to satisfy their patients' and their own needs.

> Tuesday afternoon at five o'clock every other week is the time for the hospital Utilization Committee and Surgical Program Committee to meet and make decisions about how the resources available are to be distributed. Currently, the most difficult issue has been the significant increase in orthopedic knee replacement procedures and the impending depletion of the budget allocated for prostheses. Today, the committees will have to decide what to do about this. The statistics have been collected and everyone who might be affected within the institution has had a chance to review these, comment on them and present their views.

> The Health Region Board of Directors meeting has just commenced and the Board has heard from the Chairperson of the Utilization Committee regarding the orthopedic implant issue. This appears to have significance to the global budget and resources of the Regional Health Authority. Dealing with it in view of the total resources available to the region from the Provincial Government, and all the other programs, new and old, requesting funding has the Board cringing. The task seems impossible.

It is pouring rain outside, and the members of the Provincial cabinet are in agreement that it must be some sign. The Minister of Health has just finished her presentation. It is evident that all the Health Regions in the Province are, again, substantially over budget and are indicating that there is no more to be gained in efficiency and restructuring. They are clamoring for increased funding. The community services drug plan has decided to cover two new drugs for the treatment of multiple sclerosis, and these are reported to be worthwhile for people with multiple sclerosis, but are extremely expensive. This information is no sooner out than the Ministers of Transportation and of Environment complain, and present their respective department's needs. Yes, everyone agrees that the rainy and gray day is a reflection of his or her turmoil.

From the standpoint of scale, what can we discover in these stories? Using my three questions will help with this. For example, question one asks what scale, global, communal or local, the phenomenon of concern occurs in. In story 1, George Smith is concerned with his own health, well-being, and life aspirations. This is clearly a local scale phenomenon. The orthopedic surgeon in story 2 has his own concerns regarding the patient he is currently operating on, but also has broader concerns regarding the welfare of many other patients he has the privilege to care for. This too is a local phenomenon, but it is different from that of the first story. There is a broadening of the circle of moral consideration that represents a broader scale than the one in story one. This is the point of my second question, which asks what are the contextual scales and what the constituent scales. In stories 1 and 2, both are considered local scale phenomena as noted above, but a difference in them is evident in the broader concerns evident in story 2. In answering question two, we can assign story 1 to a constituent scale and story 2, because of its broader concerns, concerns that include those of story 1, to a contextual scale in relation to story 1. Therefore, questions one and two allow us to describe the scalar nature of story 1 as local-constituent and that of story 2 as local-contextual. Story 1 is, in effect, a part of a broader story. This is true for each of the stories in turn. In as much as the response rate for each of these may be slower the wider the circle of consideration, the broader, slower-rated scales are considered contextual for the more local, faster rated scales. The point here is that even within a single major scale category, there is room and reason to identify further boundaries that allow the description of parts and wholes, that is, constituents and contexts. The third of my three questions directs our inquiry into the integrity and good functioning of the contextual scale. For example, in stories 1 and 2 the integrity and good functioning of the orthopedic surgeon may relate to his ability to see and care for only so many patients before he is "overloaded" and begins to experience a decreasing quality of care to the patients he is caring for currently (perhaps less time dedicated to each) and / or personal health problems which then impact on his ability to care for patients both current and new. When such impact on the context is evident or close at hand, the contextual scale may legitimately place constraints on the local scale. In this scenario, for example, such constraints may include strictly controlled access to consultation and surgery or more time to himself for renewal, both of which may increase waiting time for consultation and treatment. In turn, the orthopedic surgeon in story 2 can relate his concerns to the contextual scales under which he functions suggesting that to alleviate his problems he needs more prostheses and perhaps another orthopedic surgeon. Using the three question framework that I have suggested and the concept of scale enables an exploration of wholes and parts and their interrelationships. Having done this the values embedded in the various scales involved are made explicit and guidance in choosing among them is fostered by the prescriptive nature of scalar concern. This is elaborated in detail in the next section.

Attention to scale, as noted above, is explicitly perspectival in nature. It seeks the breadth and depth of perspectives available in each case. In this, attention to scale can be seen to be compatible with Aristotelian phronesis (practical reasoning) aiming to recognize and strike a balance between and among the full particulars of a whole human life, "acknowledging human participation at many levels in the structure of reality" (Emberley, 1995, p. 20). Getting involved is exactly what emphasis on scale (including their accompanying constraints and values) encourages us to do. Furthermore, it clearly leads one to the ethical and moral work envisaged by Emberley (1995). Moral work where "the individual acts to reproduce in himself the substance presumed to have ultimate reality because it is constant and durable ... [where] ... to 'be' ethical or moral is to 'have' communion with what is taken to be ultimately real" (Emberley, 1995, p. 31). Emphasis on relationships, inter-relationships and interdependencies, and our responsiveness and response to them is of central importance here. Scalar consideration, in this view, also shares features with a care ethic where relationship and response are of core interest (Gilligan, 1982; Self and Skeel, 1992).

From a descriptive point of view, the concept of scale in the realm of bioethics has a great deal to offer. It demands the explicit identification of the parts and wholes, and requires an exploration of the nature of the relationships, and dependencies, and interdependencies of the parts and the explicit consideration of the characteristics of the contextual scale (whole).

Prescription: Moral Decision-Making

Kluge (1992) stated that "ethical decision-making is a matter of psychology." Specifically, the process by which a moral decision is reached involves a "psycho-social integration" of socialized, internalized, cognitive concepts that provide the basis upon which moral decisions are made. Citing Kohlberg's (1981) formal cognitive approach to moral development and moral reasoning, Kluge limited the psychology to rationalism while others would include some role for emotivism in moral reasoning. Either way, there is an understood psychological underpinning to our moral reasoning that can be assessed and analyzed (Gorman, 1989; Self & Skeel, 1992). Drawing on work from Piaget and others, Kohlberg (1992) has developed a theory of moral development that he claims is both philosophical and psychological. It consists of culturally universal stages of moral development that are invariant in their order and are qualitatively based in cognitive theory (Dienhart, 1982; Rich & de Vitis, 1985). This progression of stages is characterized as moving from a centered, superficial perspective where judgment is based on immediate perceptions of circumstances, to a decentered, more profound perspective where judgment is based on a broader understanding of the inferred or underlying meaning (Gibbs, 1991) embedded in the circumstances. Rich and de Vitis (1985) suggested that Kohlberg sees morality as "a unique realm that has distinctive formal characteristics in formulating moral judgment: impersonality, universalizability, ideality, preemptiveness ..." (p. 93). That is to say, morally mature thinking involves impartial, universal, idealized rules or principles that are normative for behavior. The end point, in this view of ethics, is one where mature moral development is constituted by a principled approach focusing on the precepts of justice, fairness, and rights. Interestingly, for a while Kohlberg entertained the notion of an ultimate stage of moral reasoning characterized by an integration of principled thinking into greater sensitivity to circumstances and specifics, a more contextual ethics. However, this stage was eventually discarded as it was not fully supported by the model from Piaget in which Kohlberg was embedded. As an aside, this gives some insight into the tensions present between science and philosophy and the workings of scientific paradigms.

Gilligan (1982), observing the lack of female participants in previous studies (not only those of Kohlberg but most of the other writers and researchers in moral development), challenged what she saw as a bias at the theoretical level of previous studies in moral development. In her own studies, Gilligan found "that women's language is substantially one of caring and interpersonal responsibility ..." that contrasts with the language of men that "... seems to be framed in terms of individual rights, liberties, duties ..." (Gilligan, 1982, p. 117). She identified this ethic of care as a storehouse of knowledge of human relationships, which has evolved "around a central insight, that self and other are interdependent" (Gilligan, 1982, p. 74).

These two ideas of moral development are not without their critics. Puka (1991), in his probing of the care-justice debate, suggests that Kohlberg's and Gilligan's approaches are fundamentally different at the level of theory development. Similarly, Brown and Tappan (1991) point out that there are fundamental differences between these two approaches regarding the assumptions about what is to be considered as evidence for moral development. Hogan (1975), using Ischheisser's terms "views in principle" and "views in facts," distinguishes a person's role structure and character structure. Views in principle are those about how one thinks one would behave or ought to behave if confronted with a certain dilemma. Views in facts, on the other hand, are generally not known until after the action in response to the dilemma has occurred. That is to say, despite what we may hold in principle, we do not really know how we would behave in a certain situation until confronted and have acted. Hogan (1975) suggests that views in principle reflect one's role structure, while views in fact reflect one's character structure. With respect to the nature of moral development outlined above, it seems that this distinction between views in principle and views in fact is shared by Kohlberg's and Gilligan's theses. That is, Kohlberg's emphasizes views in principle and Gilligan's emphasizes views in facts. This controversy in moral psychology continues. However, given that both conceptualizations, justice and care, are part of everyday living, it seems fair to say that both perspectives are valid and that the true nature of moral psychology and decision-making most probably incorporates features of both.

Blum (1988) argues that a care perspective is important in knowing how to apply certain principles in a certain situation when specific particular concerns exist that would not be considered within the universal principle itself. For example, consider the principle "First of all, do no harm" frequently cited in medicine. In order to realize this principle, the doctor looking after Mrs. MacDonald must know or come to know many specific and particular things about Mrs. MacDonald and her circumstances to be able to know what, for her, constitutes harm. The principle itself does not tell us what constitutes harm. There is a similar controversy between the universal and the particular in moral philosophy, particularly in the newer discipline of bioethics. Toulmin (1981), in his paper "The Tyranny of Principles," referred to these two views of moral orientation, justice and care, as the "ethics of strangers" and the "ethics of intimacy," respectively. He wrote, "in the ethics of strangers, respect for rules is all, and the opportunities for discretion are few. In the ethics of intimacy, discretion is all, and the relevance of strict rules is minimal" (p. 35). Thus, there is a recognized concern with the adequacy of a solely principled approach to moral deliberation. Alternative and complementary approaches emphasize the role of moral emotions and the importance of context, particulars, and relationships. Scale is one such complementary approach.

From this brief consideration of aspects of moral psychology and moral reasoning, I would like to explore some aspects of ethical theory, specifically prescriptive or normative ethics (Appelbaum & Lawton, 1990; Beauchamp & Childress, 1994; Beauchamp & Walters, 1994; Fletcher, 1966; Kluge, 1992; MacIntyre, 1984; Mappes & Zembaty, 1992; Raphael, 1981; Regan, 1986; Roy et al., 1994; Sahakian, 1968; Trusted, 1987). My discussion here will focus on an understanding, taken predominantly from bioethics literature, of the meaning of prescriptive statements (norms) in two broad categories of normative ethics that I will refer to as principlism and contextualism.

Deontology and Utilitarian Teleology are two generally accepted principle-based normative theories. They have developed, over the past two hundred years, within a worldview, dominated by scientific positivism and philosophical rationalism, that sees "genuine knowledge" as that acquired through reason and that this knowledge represents the truth about the world (Raphael, 1981, p. 18). The accepted method of this school is logical deduction from sets of principles, and an ethical defense of a moral decision is one of justification based on these principles.

Recent re-considerations of normative ethics and the mode of defending a moral decision, arising from a post-positivist foundation, have de-emphasized the centrality of principles and have elaborated an increased emphasis on context and relationship. The method of this practical ethics is dialogue, and ethical defense is more concerned with explication and understanding than with justification based on principles alone (Hoffmaster, 1990). A scalar approach to issues in bioethics captures this later view quite directly with its emphasis on context, parts, and wholes.

Scale and Guidance in Values Priority

In any ethical issue, the analysis leads to a judgment about what is to be considered as mandatory, permissible, tolerable, or prohibited (Roy et al., 1994, p. 28).

The process of this practical wisdom involves the asking of, and the attempt to answer, the question "Which general norms for guidance and evaluation of conduct are worthy of moral acceptance and for what reasons?" (Beauchamp & Childress, 1994, p. 4).

As noted earlier, in hierarchy theory a lower frequency (slow rated) process is considered to be a higher-level system because it operates on a larger and longer spatiotemporal scale. It thereby provides a context perceived by the higher frequency (faster rated) scales as stable and in which the higher frequency systems can unfold. That is, lower frequency processes set limits or constraints for high frequency processes, and the slow rate of the contextual system effectively dampens and filters the activities of the lower level higher frequency activities. However, this dampening and filtering effect can be overwhelmed if the cumulative activities of the lower level systems create perturbations that extend beyond the rate capabilities of the contextual system to respond; that is, the limits or constraints of the contextual scale are broken or overwhelmed. In this circumstance the lower level systems will experience strife that will result in their return to function within the constraints of the higher-level scale, or the higher contextual scale will fail and change to a new form. This new level of system function will be at a lower level than the previous one and may no longer support the lower level systems that previously existed and they too are forced to change. Such changes are most often negative with diminished diversity in the systems and less complexity and creativity (Allen & Starr, 1982; Wilby, 1994).

As a simplified ecological example let's consider the life cycle of a forest insect. This may be several weeks or a few months during which time it may devour a certain volume of forest biomass over a given space (area). The life cycle (rate of activity) of the forest, though, is measured in centuries over an area that is immense. From the insect's perspective, the forest appears perfectly stable and unchanging, accommodating the effects of the insect, damping or filtering them out as if they never occurred. However, if the individual insect becomes a huge infestation, their combined activity, even though they have a rapid rate life cycle, may overwhelm the ability of the lower rate forest cycle to accommodate the resulting effects (loss of biomass). The limits (ability to regenerate biomass) tolerated by the forest become evident to the insect infestation through the loss of nutrient supply, and their numbers decline. Ecologically, such a limit or constraint is reflected as the "carrying capacity" of the forest system. That is, the limits determine the insect population size.

The recognition of limits or constraints on the parts by the whole is, in the context of the human moral agent, an ethical or moral responsibility to limit behavior. Such control, constraint, and limitation of behavior is one of the four fundamental moral sentiments (self-control) elaborated by Wilson (1993) and Crigger (1996). Aldo Leopold (1949/1970), in his development of "the land ethic" (p. 237), has also stated that: "An ethic, ecologically, is a limitation on freedom of action in the struggle for existence" (p. 238). Similarly, in human social systems, an individual life unfolds and decisions are made in a time-scale of days to the few years of one human life span. Society, however, operates on a larger and longer spatiotemporal scale and is, therefore, perceived by the individual human as a relatively static context in which one lives out one's life. Society, as the slow-rated contextual level for the individuals that compose it, sets constraints or limits (e.g., laws, rules and regulations) on individual behavior. Giampietro (1994) put it this way, "the lower frequency changes in behaviour of society means that decisions made at the level of society are perceived as laws when read from the individual level. In this way, individual behaviour is affected by societal behaviour in the form of a set of constraints defining what the individual can and cannot do" (p. 617).

As noted earlier, the disciplines of ethics and, specifically in the context of this work, bioethics also deal with constraints to action and behavior. However, so far, these disciplines have been primarily concerned with inter-human, inter-personal systems. Only in recent decades has there been renewed and increasing interest in viewing constraints to human behavior that come from outside humanity itself as ones that humanity should legitimately consider, comprehend, and embrace in moral terms. This is clearly suggested by Rolston (1988), who states: "Interhuman ethics has spent the last two millennia waking up to human dignity. As we turn to a new millenium [sic], environmental ethics invites awakening to the greater story of which humans are a consummate part" (p. 338). The concept of scale from hierarchy theory brings to the fore the age-old philosophical conundrum of humanity's place in nature and nature's place in humanity. The part-whole dialectic is consistently addressed by an explicit emphasis on scale.

I wish to re-emphasize the description of scale outlined in chapters 2 and 3. In every issue considered, explicit attention to scale requires this part/whole dialectic to be addressed. From a human perspective, this means we must confront our humanity as both a whole and as a part of a greater "whole." Hierarchy Theory, as a systems theory, concentrates on the interactions and interrelationships between objects rather than on the objects themselves (Haigh, 1987). Attention to scale in hierarchy theory emphasizes the concept that each component of the hierarchy is seen to have a dual nature, a concept that Koestler (1967, 1978) has referred to as a "holon." Each holon, therefore, is always facing two conflicting values, that is, maintaining the integrity of self as a whole, yet necessarily recognizing itself as a part (subwhole) of the contextual scale (whole) within which it develops and finds its meaning and upon which it is fully dependent. These values imply different courses of action and behavior, are clearly scalar in nature, and are, just as clearly, of central ethical concern. It is necessary to have a clear conceptualization of the nature of the interactions and interrelationships between and among the various components of systems in any hierarchy. Haigh (1987) pointed out that "Hierarchies are synergistic open systems whose critical aspects are in the balance between the self-assertive and the integrative tendencies of their member holons" (p.190). Sahtouris (1989) referred to this process of coming to grips with our whole-part nature and realizing a balance as establishing "mutual consistency" (p. 34):

> Any holon containing smaller holons ~ such as the earth full of bacteria or a body made of cells ~ tempers the individual autonomy of its components with its *own* autonomy, which is *their holonomy* [italics in original]. Any individual human, for example, must transcend simple self-rule and integrate him- or herself with the rules of society, while human society must transcend its autonomy and integrate itself with the holonomy imposed by the autonomy of the planet. The balance between any holon's autonomy and holonomy must be worked out as mutual consistency if the holon is to survive as part of a holarchy—and it cannot survive in any other way. (Sahtouris, 1988, pp. 54–55)

Here, autonomy is to be understood as the actualization of the natural tendencies of the unit under consideration rather than the ethical understanding of autonomy as self-determination and free choice.

One way to help understand the perspective from which this concept is coming from is to adopt Buber's (1922/1970) "basic word I-You" (Kaufman, 1970, p. 53) and personify the "other" in the natural world as elaborated by Tallmadge (1981). If we then, conceptually, personify each holon, we can say that the dual tendency of the holon establishes two levels of responsibility from which it is required to act:

1. To meet its own needs and maintain its own integrity (that is—as a whole to allow for the good functioning of its parts allowing each to perform/unfold to its optimum level) yet;
2. To respect the constraints of the whole of which it is a part so as to maintain the integrity and good functioning of its contextual whole.

With regard to these dual responsibilities, one might ask, why a holon should limit itself when it could do just as well or better with a different whole? The difficulty here for the holon is that it can not know for certain that it can, indeed, do better with a different whole, particularly when the long term is considered. Since, by its very nature, the integrity and good functioning of the whole includes its constituents, a broadened view of self-interest, along with simple prudence, suggests the validity of these responsibilities.

Giampietro (1994) explained this dual responsibility in this way: "What enforces the individuality of the holon will reduce the stability of the higher level. On the other hand, an excessive control from the higher level of the freedom of the lower level holons will endanger the stability of those lower level holons as mentors of the hierarchy" (p. 618). Taking an hierarchical, scalar perspective, Giampietro (1994) went on to describe the nature of this mutual consistency:

> In terms of hierarchy theory, vertical coupling is defined as the interaction between different levels, and horizontal coupling as the interaction among holons of the same level. A healthy hierarchy has to reach a compromise between the intensities of both horizontal and vertical coupling (not too loose and not too strong) to avoid the destruction of its complexity. This means that the values expressed by each of the two perspectives are important, and that they have to be considered at the same time in a holistic perspective, which should refer to the "goal" and "meaning" of the hierarchy as a whole. The stability of each level depends on the stability of the other levels, either higher or lower in the hierarchy. This concept is referred to as 'double asymmetry' or the need for "equipollence." (p. 618)

Complex systems are open to description through reductionist analysis ("near-decomposability" (Simon, 1962)) where the perspective from which the analysis takes place is hierarchically a clear statement of the spatiotemporal scale chosen for analysis, and which effectively isolates the chosen spatiotemporal scale from higher and lower level processes or behaviors. This means:

> In order to have a description that is effective in dealing with a particular problem: a) The choice of the time and space scale should be compatible with the parameters describing the dynamics we want to study; b) The behaviour of the higher level, observed from the particular spatiotemporal scale chosen, should ... [be] ... so slow that it can be considered negligible [stable]. In this case, the higher level can be accounted for in the description as a set of external constraints or boundary conditions; and c) The perturbations generated by the behaviour of the lower level should be negligible as well. The lower level "noise" is eliminated by a statistical (macroscopic) description of events at the lower level. (Giampietro, 1994, p. 619)

To bring this discussion into the moral domain, the nature of hierarchical systems, as outlined above, means that from the standpoint of a values description:

> A spatiotemporal scale must be identified and stated for the values that are identified for description or consideration (global, communal, individual from the perspective of the framework outlined earlier) and;
>
> The values of the contextual system within which this selected spatiotemporal scale is embedded must be explicitly identified and recognized as stable constraints on the chosen spatiotemporal scale (distinguishing context and constituents as identified in the framework outlined earlier).

From the standpoint of humanity, we are at one and the same time the observers and the observed, both wholes and parts. Since we have the innate capabilities of reason and foresight enabling us to be aware of this dialectic, it is incumbent upon us to recognize our place in the "holarchy" (Koestler, 1967, 1978). It is necessary to recognize and be explicit about the various spatiotemporal scales in which we as individuals, family, community, society, and species operate, and to use this practical wisdom in a morally sound way.

It seems evident to me that present ethical systems have recognized the individual and societal spatiotemporal scales. The normative and justificatory aspects of these ethical systems seek to guide and explain behavior at these scales. However, they have not viewed these in the hierarchical, scalar terms, as I have outlined them, and this has resulted in cross-scalar (vertical coupling) arguments that fail to be clear on the priorities of constraints and limits that contextual scales place on their constituent subscales. Values dialogue between subscales and the contextual scale needs to explicitly recognize the nature of this vertical coupling. Furthermore, in my view as well as that of others (Donnelley, 1998; Giampietro, 1994; Potter, 1971, 1988, 1999), the spatiotemporal scale that places humanity within ecologic and biospheric systems is one whose principles and values we have yet to fully integrate into our moral systems. For instance, Giampietro (1994) pointed out that hierarchy theory is useful in understanding the concept of sustainability precisely because of its emphasis on scalar values:

> Humans are now facing this problem of inability to describe the behaviour of the system as a whole. Ecological analyses are not able to deal with human behaviour and economic analyses ignore the importance of natural processes. Humans and their economic systems have expanded so much as holons that they are endangering the stability of the higher level (ecological system) of the hierarchy to which they belong. This implies that when technological development of socioeconomic systems is assessed, the

biosphere level and its related perspective must be included in the process of comparing and mediating between contrasting values. (Giampietro, 1994, pp. 620–621)

Explicit attention to scale provides prescriptive guidance in these issues by directing the support of contextual scales, the values of which are the constraining limits or boundaries for the actions of lower level scales in the hierarchy. The greater whole, contextual scale, of which the human species is a part is the biosphere. The normative nature of the concept of scale requires that moral weight be given to the values and principles of interaction that are operational at this biospheric scale. The concept of scale with its normative nature becomes a useful addition to bioethical analysis and decision-making because:

> When crossing different timescales the operationality of many traditional ethical principles blurs, since the application of ethical rules depends on implicit definition of an operational timescale.... Hence, the high level of uncertainty faced when dealing with intergenerational analysis means that theoretical principles of ethics are of little help in providing practical solutions. (Giampietro, 1994, p. 622)

Similar to the earlier example of the principle "first of all, do no harm," where knowledge about the particulars of a patient is needed, my framework of questions and scales directly seeks out these particulars and, as elaborated, provides prescriptive guidance. From the standpoint of intergenerational concerns, since these are communal-scale phenomena, the values inherent in their integrity and good functioning would generally have priority over those of local scale phenomena.

Westra (1994) also makes a "normative appeal" to the "integrity of the whole." It seems to me, too, that "there is normative appeal in a system which will make valuation depend, in part, on the presence of properties which cannot survive the disruption-restoration process" (Westra, 1994, p. 57) that occurs when higher system constraints are overwhelmed by lower system activity. A broadening of bioethics to incorporate the concept of scale, as it relates to recognizing contextual and constituent scales and the prescription to emphasize the higher level contextual scale, provides a different way of viewing and understanding the complex nature of intergenerational and multiscalar issues than we are familiar with in current bioethical analysis.

Human Nature and the Capacity of Foresight

Another area that I think it is critical for us to consider because it bears on how humans act and react, is the nature of human nature. I am particularly referring here to the nature of our (human) moral life. I believe, along with others (Crigger, 1996; Potter, 1990; Wilson, 1978, 1993) that to a greater rather than lesser extent, our moral life is biologically mandated. Following Wilson (1993), Crigger (1996) explained that "our natural moral sense encompasses four fundamental moral sentiments (she concedes there may be more) that have (had) adaptive value for the human species: sympathy, fairness, self-control, and duty" (p. 33). These are all rooted in our social nature and are nurtured and reinforced during socialization and the development of the self. Further, coming from Tierney (1994), Crigger noted the importance of our ability to imagine the good life. This moral imagination, Crigger (1996) pointed out, is the goal of "developing a theory of imaginative idealization as an account of how concrete moral representations enter the self" (p. 34). In a similar way, supported by Potter (1990), our human capability of foresight has a biological proclivity and adaptive value that carries with it the duty to respond, not only prudentially, but also morally. Having the capability of anticipating some good outcome by a certain course of action, or of preventing some harm by a certain course of action, clearly integrates our response with the other moral sentiments mentioned earlier. Furthermore, not using our ability of foresight to its optimum is an impoverishment of our humanity, a failure to achieve excellence teleologically in Aristotelian terms. This emphasis on foresight is important because it is at the basis of Potter's plea for a bioethics that considers the long term as much or more than the short term. In fact, Potter (1990) characterized humanity's tendency to responsiveness in the short term as our evolutionary "fatal flaw" and wondered if a global bioethics (as he called his book) can overcome it. A view to the long term and responsiveness to it is also a major component of scale as I have elucidated it. The explicit identification of scales, and especially the contextual wholes, by its very nature takes us into a long-term perspective and enables the assessment of the critical supportive or foundational values of the contextual scale.

The Proposed Framework and Values Priority

The attention to scale, parts, and wholes, suggested here is perhaps most useful in policy considerations. This approach is conceptually different from current approaches to bioethical analysis and decision-making. Indeed, it is a valuable addition to these current approaches rather than an alternative. The scalar, parts, and wholes, approach strives to explicitly identify varying scales and seeks the nature of relationships and the set of values associated with these different scales. It uncovers the fundamental values of each scale and gives guidance as to the constraints set by each scale. The guidance is this: one must not undermine the integrity and the good functioning of one's contextual scale; yet contextual scales must allow constituents the ability to reach their own optimum level of integrity and good functioning. Following from the view of scale expressed in chapter 2, when deliberating on values, explicit attention to scale means:

1. that a scale must be identified and stated for the values that are being considered, and
2. that the value(s) integral to the contextual scale within which the selected constituent scale is embedded must be explicitly identified and must be recognized as constraints on actions undertaken in the constituent scale.

Let us look at each of these and understand what they require. The first statement accepts that our values are scaled. That is to say, as presented in chapter 2, some values have more currency at one scale than others (see also Norton, 1995; Norton & Ulanowicz, 1992). This being the case, it is necessary to qualify the values under consideration by identifying the scale in which the value is active. The second statement has two components that were presented in detail in chapter 2 and the first two sections of this chapter, that is, a descriptive component and a prescriptive component. The descriptive aspect requires the explicit identification of the part(s) and whole. In any issue, it is necessary to be explicit about which scale is the contextual one, the greater whole, and which the constituent one(s). The prescriptive aspect requires that pre-eminence be given to the values of the contextual scale by recognizing them as limitations or constraints under which the values and activities of the constituent scales are free to unfold. It is very important to understand that this does not imply a fascist-like domination of the part(s) by the whole. What is necessary is an acceptable balance, or using Sahtouris' (1989) metaphor, a "mutual consistency" between / among the scales. This, not so simply, means a "working out" of what supports the integrity and good functioning of each given the constraints involved.

There are, I think, at least six reasons to support the claim that the set of values that protect and promote the integrity and good functioning of the contextual scale should take precedent over the set of values of the constituent scale(s).

1. The good of the constituent(s), particularly the long-term good, is entirely dependent on the integrity of its context. This point has two aspects.

    (a) By definition, the context subsumes the constituents, therefore, protecting and promoting the integrity and good working of the context, by its very nature, carries with it the promotion of the good working of its constituent(s). That is, we can say, all things considered and especially over the long-term, that what is good for the whole is good for its parts. However, we cannot say that what is good for a single part is good for the other parts or for the whole.

    (b) In general, each part is best served when governing itself, that is, when seeking its self-interest. The conceptualization of scale that I am proposing does not preempt this, but does demand from the constituents a grander understanding of self-interest. When the constituent understands and appreciates that its good, especially its long-term good, is intimately related to the integrity and good functioning of its context it is evident that its good functioning may need to be an optimum, rather than a maximum. This is particularly so when the integrity and good functioning of the context is deemed to be at risk. In other circumstances when such risk is not suspected or when it certainly is not present, the constituent proceeds on the basis of its best interests reiteratively checking the status of the context. Furthermore, the self-interest of the context carries with it the good working of its constituents, so it is in the interest of the context to be as minimally constraining as its overall integrity allows. This cannot be said in any way to be riding roughshod over the values of the individual parts, particularly as viewed over the long-term. Clearly, acting in such a way as to protect and promote the integrity and well-being of both the contextual and constituent scales is preferable to the alternative of sacrificing one for the sake of the other. This is the 'working out' referred to above.

2. As elaborated earlier (chapter 2) in the introduction to scale as taken from hierarchy theory in ecology, by its very nature the contextual scale is more stable than the constituent scales. It is less affected

by perturbation. Such stability is both a short-term and long-term value and is dependent upon the integrity and good functioning of the context. Constituents benefit from existing in a stable context where perturbations are less frequent.

3. Furthermore, as pointed out in chapter 2, in general, if the context is challenged sufficiently to make it unstable and cause it to change, the short-term effect (the one that will most affect the constituents) is a reduction of context to a lower scale. This lower scale context will have less diversity, less stability and less ability to protect its constituents from further perturbation. It is to be expected that constituents, too, will suffer. Although some may flourish over the short-term with the loss of constraints, many will be lost or themselves diminished and it is generally expected to take a long time for the context to regain stability and diversity comparable to the previous context. However, it is to be noted that we are not talking about static systems. Changes in context do and are occurring constantly, but are slow long-term changes that allow constituents to adapt as required.

4. Also, from the earlier discussion in chapter 2, it is evident that diversity is best protected and promoted, maintained and enhanced, or at the very least not diminished, when the integrity and good functioning of the context is maintained. In general, if the context is disrupted and diminished, diversity is also diminished. Diversity is here taken as both a short-term and long-term value. The greater the diversity the greater the stability of the context in face of perturbations and the more the constituents are protected. Furthermore, this diversity is a direct reflection of the quantity, creativity, and the integrity and good functioning of the constituents themselves. This has to be considered a good thing from the standpoint of the constituents as well as the context.

5. By its very nature, priority support for the integrity of the contextual scale necessitates a long-term perspective as opposed to a short-term perspective and thereby counters what Potter referred to as humanity's fatal flaw mentioned in the preceding section. Attention to scale, as I am proposing, naturally counters the tendency to discount the future and long-term values. Speculatively, this focus on short-term gain at the expense of the long-term is understood as an evolutionary advantage to early man, and an expansion of this to a full utilization of human potential for foresight (preeminence of long-term over the short-term) to be our evolutionary future.

6. Finally, by its very nature, a prescriptive focus on the constituent is exclusive. On the other hand, by its very nature, a prescriptive focus on the context is inclusive. This inclusiveness is a common theme in spiritual and religious discourse and can be interpreted to be akin to love (Vanier, 1998).

It is not my intension here to introduce any conceptions of intrinsic value as the whole cannot exist without its parts nor can the parts exist without the whole. This argument stands alone without any appeal to intrinsic value, although it may well be for some that love presupposes such valuation. In my view, love is more a recognition of the connectedness of all things, while separation, the lack of such a sense of connection, is the state of exclusion. Ecological and evolutionary sciences, let alone religious and spiritual thought, speak to the veracity of this interconnectedness.

You will note that points 2, 3 and 4 give evidence for important aspects of the context and directly support the claim made in point 1. Point 5 presents good reason to emphasize the long-term in our deliberations and not only stands alone as an argument for priority being given to the contextual scale, but also directly supports the claim in 1 with its emphasis on the long-term.

It should also be evident that the sixth point follows from the others. It emphasizes the notion that the context includes the constituent(s) and that the well-being of both is integrally connected. A loving response takes this notion seriously.

One important critique here relates to the assumption that what is good for the whole is good for the part(s). One can imagine all sorts of scenarios where the constituent suffers when constraining what it sees as its best interest, for example the use of a pastoral commons. In this scenario, since each herdsman gets the return on his cattle without paying for the commons, he maximizes his profit by adding more head to the herd, yet eventually the carrying capacity of the commons enforces the constraint and all the herdsmen suffer, though the one who added the most cattle may suffer less. A broadened view of self-interest here may preserve the herds and the commons indefinitely, and therefore the potential for profit, but the profit is optimal rather than maximal. Ultimately, acceptance of this view of whole and part may rest on the degree to which one sees oneself connected to other, past, present and future. If this sense of connection, which must be felt as much or more than cognitively perceived, is strong then the presumption that what is good for the whole is good for the part will ring true. If it is not so strong, even though such separation flies in the face of ecological knowledge, then it will be questioned until proven, but by then it may be too late.

Consider also the scenario of a shipwreck and a lifeboat overfull with survivors. It is clear that if someone is not let go then all will succumb. Self-interest, understood as survival, will likely result in chaos and murder. This solves the problem, though this behavior cannot be condoned. A broader view of self-interest, however, could lead the eldest of the survivors, or the most injured for example, to choose to let go, allowing the survival of the whole group and the recording in history of the sacrifice made. Such stories are common in our history extending back for millennia.

The implication of this line of argument might seem to be that there is never any real conflict between values at different scales, just a narrow understanding of what constitutes best interest. In ecological and evolutionary terms, constituents do, at times, get sacrificed, that is, disappear from an ecosystem or become extinct. This would seem to be a state of conflict. However, such occurrences generally represent the failure of the constituent (organism/species) to maintain or to establish its "fit" in the context, whereby it is lost. This failure to maintain or establish "fitness" can be framed as a narrow view of best interest, supporting the notion that true interscalar (vertical) conflict may not exist. The process of establishing "fitness" may be framed in terms of a working out of a mutual understanding of what constitutes best interest for both the context and constituent. However, this process of "working out" fitness is not without its struggles and, in this view can be taken as a form of conflict, so I think it is still reasonable to speak in terms of interscalar conflicts understood as the struggle around best interest over the short- and long-terms. A state of mutual consistency exists when there is a convergence of understanding of best interest so that the integrity and good functioning of both contextual and constituent scales are achieved. Intrascalar (horizontal) conflict exists between or among constituents as each attempts to define its individual "fit" along with the others using what capabilities each has.

Despite this apparent denial of conflict between scales, as pointed out above I think it reasonable to consider the struggle in establishing mutual consistency as a form of conflict. Certainly, controversy over what values count and conflict between values is no less evident when scale has been identified than when scale has not been explicitly elaborated. However, taking this issue of scale seriously and explicitly identifying context, constituents, and state of contextual integrity helps discernment among values and gives a basis for working out these conflicts and controversies.

Also, with respect to this idea of constraints, the necessity to establish and maintain an ecological / evolutionary "fit", what I believe Sahtouris' (1989) meant by her term "mutual consistency", may require some constraining of the constituent's development and activity. This usually involves various feedback loops, so that optimums (we might consider these the carrying capacities ecologically) rather than maximums in development are realized. This limitation on the actions of a constituent is not simply an imposition on the constituent but is, in fact, a limitation that is in the constituent's own interest and that is there for it to discover and come to understand. This notion bears some similarity to the game theory paradox of the prisoner's dilemma. Here, choosing a mutual cooperation strategy involving a constraining of one's maximum utility, rather than choosing an individual strategy, results in an optimum utility that is shared by all. That is, people who constrain maximization of their utility, "constrained maximizers" (Gauthier, 1985), appear to do better overall, in terms of their utility, and have wider opportunities for interaction than people who do not act with such constraint.

Given the discussion above, there is reason to take this notion of scale seriously and I propose that a framework of questions could be used to address its requirements. This framework would help uncover the contextual (whole) and constituent (part) character of the scales as well as the values pertinent to each. Prescriptive guidance is then available about what actions must be considered mandatory, what actions permissible, what only tolerated, and what actions prohibited based on the pre-eminence given to the contextual (whole) scale and the desire to achieve a mutual consistency between the contextual and constituent scales. For any specific circumstance the set of questions outlined above, in the section on the proposed framework of scales, can be asked to elaborate the scales, their character, wholes and parts (context and constituents) and their respective values:

> What scale, global, communal or local, is the phenomenon (action, object or value) of concern occurring in?
> What is (are) the context(s) and what the constituents?
> Is the integrity of the contextual scale at risk?

Together, these three questions address both of the aspects of scale discussed at length in this chapter. All three questions seek a scalar description of the issue under consideration. The first two questions enable the identification of scales, wholes and parts, and their respective sets of pertinent values, and opens up the relationship between or among them for scrutiny. The third question is focused on both description and prescription. It requires the moral agent to assess the status of both the context and the constituents particularly with regard to their integrity and good functioning. This is not a single descriptive step, but is intended to be an iterative process constantly being re-addressed. When it is discovered, or suspected, that the integrity and good functioning of the context is at risk or that its status is uncertain, question three reminds us of the prescriptive nature of scale, as presented and argued above. The moral agent is guided to the support of the value set of the context as a limitation on the value set of the constituents if the intention is to maintain contextual integrity and good functioning. This allows the moral agent to be imaginative in seeking to establish a state of mutual consistency where constraints are recognized and honored, while the integrity and good functioning of both the contextual and constituent scales are assured. This is the desired state in which there has been a convergence of understanding of the nature of best interest for both the contextual scale and the constituent scale.

· 4 ·

CASE STUDY: AN EXAMPLE OF SCALE

Forestry and Clear-Cutting

The clear-cutting controversy has a lengthy history with roots in the exploitation of the Canadian forest dating back to the earliest days of European settlement. The history of land use worldwide bears witness to these words of Francois Reese de Chauteaubriand, "Forests precede civilization and deserts follow them" (quoted. in Harrington, 1990, p. 34). North and East Africa's once expansive forests have been aggressively cut for fuel, fiber, and crop land, and deserts are now the dominant landscape accompanied by tremendous human suffering. Similarly, the great cedar forests of the fertile crescent (the Cedars of Lebanon) are long gone (Eisenberg, 1998). We are now generally aware of the extreme importance of our forests, landscapes, and ecosystems to the health of the land and the planet. Forests worldwide play critical roles in the capture of energy from the sun by photosynthesis, the protection of soil from erosion by interruption of rainfall, uptake, storage, and slow release of moisture and fixation of soil by root systems and thus the protection of our watersheds from siltation. They are intimately involved in nutrient cycles and fixation of the nitrogen needed for protein synthesis by all living things. They purify the air and modify the climate through processes of respiration and evapo-transpiration. Forests function as huge carbon sinks, storing carbon and helping to maintain low atmospheric carbon dioxide levels. Finally (but not least), they are home to a myriad of living things providing macro and micro habitats for all forms of life from protists and fungi to flowers, insects, birds, deer, and humans to name a few. The recognition of these life-supporting activities of forest systems has come to us through the science of ecology and may be referred to as "ecological values." In contrast, human use of the world's forests as a resource for human material needs, fuel, fiber, crop, and grazing land has not generally recognized the ecological values of forests despite the insights of many notable individuals and groups over millennia. For instance, Plato, recognizing the forests' role in maintaining pure water (Harrington, 1990), advised measures to control deforestation. Franklin D. Roosevelt is reported to have observed that forests are "the lungs of our land, purifying our air and giving fresh strength to our people" (quoted in Harrington, 1990, p. 77).

In North America, the vast eastern forests have been steadily harvested, initially to clear land for human habitation and to create crop and grazing land, and then, more intensely and commercially for lumber and for pulp used for making paper. Within two hundred years of the major arrival of European settlers to North America, virtually all of the old growth forests of eastern North America have been felled. Modern forest harvesting is machine- and energy- intensive rather than labor- intensive. This technology has allowed the forestry industry in Canada, for example, to add 55 million square meters of forest to the commercial forest base per year over the ten years prior to 1994 (Government of Canada, 1994) and hastened deforestation. Since 1952 wood harvesting in Canada has increased by 50% (Harrington, 1990). Furthermore, modern forestry has resulted in a degradation of the land by its machinery and techniques (Harrington, 1990).

In Canada, over-harvesting has been a recognized problem, especially in British Columbia and the Maritimes (Harrington, 1990). Clear-cutting, the harvesting of all merchantable stems, remains the primary harvesting method accounting for 90% of Canadian wood harvest (Government of Canada, 1994). Canada's forestry practices and policies historically have been less than ecologically acceptable and have come under both internal and international criticism for a number of years. Clear-cutting, as a harvesting technique, has and continues to be, a major bone of contention in Canada. This controversy is exacerbated by the number of players involved each having his or her own particular view of the merits and demerits of clear-cutting. It has been useful to consider these various players in terms of their position in the sociopolitical system in which we all live. The conception of society's structure that is most useful here is that of Fischer (1990) as modified by Miller (1999). Both authors argued that we live in a "technocorporate state," in which power and control are arranged hierarchically (used here in the context of uni-directional command and power, rather than in the ecological sense in which I have used this word). At the top, as it were, is an iron-triangle of power composed of the political and corporate elite, served by a cadre of scientific and bureaucratic professionals. This triangle of elites controls policy-making and tends to seek its own self-preservation. At the bottom, in terms of power, rests the silent majority of relatively disenfranchised citizens, those who have little individual impact on policymaking. Within the silent majority, however, are numerous interest groups who have organized themselves into vocal associations pressing their various causes. Between the powerful and the powerless lie the media who serve to move information within the system, as well as playing no small part in shaping policy options. As we shall see below, each of the levels in the system of power appears to abide by different sets of values. Some are willing to recognize a long-term scale characterized by ecological values (communal scale in terms of this work), while others simply focus on the short-term scale, characterized by concern with current economic imperatives (local scale as I use it).

The analysis that follows makes use of a report of a major evaluation of forestry practices (including clear-cutting) in Canada undertaken by the House of Commons' Standing Committee on Natural Resources entitled *Canada: A Model Forest Nation in the Making* (Government of Canada, 1994).

"Canada: A Model Forest Nation in the Making"

The House of Commons Standing Committee on Natural Resources (hereafter referred to as the Committee), which called numerous witnesses, addressed several areas where there is significant uncertainty and lack of knowledge with regard to our silviculture methods, related technology, and their ecological effects. After considering all views, the Committee put forward its report and recommendations, which will serve here as a source of information on the extent to which matters of scale are recognized or not recognized in forest policymaking. For example, the Committee clearly pointed out the importance of the forestry industry to Canada noting that it is a "$40 billion industry ... [that] directly and indirectly generates one out of every sixteen jobs in Canada. ... [and that] ... forestry activity is an extremely important stimulant to regional development" (Government of Canada, 1994, p. 1). Similarly, the Committee recognized that forests in Canada are valued for their "contribution to the country's $26 billion tourism and recreation industry, and to hunting and trapping activity" (Government of Canada, 1994, p. 1). Also, the Committee considered annual, five-year and 25-year plans for forest management (Government of Canada, 1994). The scale, following the framework proposed and outlined above, for all these activities, values or interests is the 'local' scale. This is the scale associated with rapid rates of development and change, and incorporates individual and group activities unfolding over days to a few years or decades. Human economic activity is a prime example of this scale.

The Committee also reported on non-timber values, referred to as "environmental and social benefits" (Government of Canada, 1994, p. 1) generated by forests. Environmental benefits that it mentions include cleaning air and water, stabilizing soils, regulating local climates, providing habitats for flora and fauna, and acting as a "carbon sink" for carbon dioxide (Government of Canada, 1994, p. 1). Social benefits identified are "cultural and spiritual" values (Government of Canada, 1994, p.1). Sustainability of the forest and all of the values attached to it is another value that the Committee considered to be of prime importance (Government of Canada, 1994). Following our framework, these concerns and values are of the "communal scale," that scale in which rates of activity and development are slow and ecological in character reflecting change that occurs over several decades to hundreds of years, and that is the context within which local scales unfold. In terms of development, the Committee considered sustainability to imply "that the use by future generations of the forest resource not be jeopardized by the use of the forest in the present" (Government of Canada, 1994, p. 24). This statement is, in my view, an implicit recognition by the Committee that there are scalar concerns in the forestry management issue. Making this explicit in terms of the proposed scalar framework would be of great benefit because it clarifies what is implicit and may be overlooked. The welfare of future generations is a communal scale value. It requires activity that is oriented to the long term defined in terms of multiple human generations. Furthermore, the requirement not to jeopardize future generations is a prescriptive one that recognizes the limitations that arise from the contextual scale on the activities of its constituent scales.

In the following section, I will look at the Committee's recommendations, singly or grouped, when appropriate, from the standpoint of the three questions outlined above, to enable explicit attention to scale. I will apply the proposed framework of scales to characterize the scales involved and discuss the implications regarding policy actions. To reiterate, the three questions to

CASE STUDY: AN EXAMPLE OF SCALE

be used as a framework for explicit attention to scale are:

> What scale, global, communal or local, is the phenomenon (action, object or value) of concern occurring in?
> What is (are) the contextual scale(s) and what the constituent scale(s)?
> Is the integrity of the contextual scale at risk?

The framework of scales proposed for this will be the matrix outlined in chapter 2:

	CONSTITUENT	CONTEXT
GLOBAL	Global-constituent	Global-contextual
COMMUNAL	Communal-constituent	Communal-contextual
LOCAL	Local-constituent	Local-contextual

In the following sections, unless otherwise noted, all page references are to the research document *Canada: A Model Forest Nation in the Making* (Government of Canada, 1994).

> Recommendations 1 and 2
>
> Recommendation No. 1: That the federal government, in cooperation with the Canadian Council of Forest Ministers, assess the economic, environmental and social feasibility of implementing a Natural Forest Landscape Management system in Canada....
>
> Recommendation No. 2: That should the outcome of this assessment prove to be favourable, the federal government promote Natural Forest Landscape Management as a useful planning approach for achieving sustainable forestry. (p. xv)

In its discussion leading to these recommendations, the Committee stated that the issue of concern was "sustainable forestry, not clearcutting" (p. 23). The Committee understood the term sustainable to imply "that the use by future generations of the forest resource not be jeopardized by the use of the forest in the present" (p. 24). The Committee identified a model for forest management, the Natural Forest Landscape Management approach (p. 31), commenting that it "is designed to examine the forest from a much broader perspective than the traditional forest stand viewpoint. ... It manages on an ecosystem basis resources other than those which have a recognized economic value." (p. 31). That is, management is "undertaken at a broad landscape level" (p. 31).

The first of my analytical questions asks in what scale the phenomenon of concern is occurring. This requires the clear definition of the phenomenon of concern, then the characterization of it as taking place in the local, communal or global scale. The primary action that the Committee is addressing is the forestry business, that is, the cutting down of trees and forests for economic benefit. The broader 'whole' within which this activity takes place is the forest landscape/community ecosystem. Following Norton's scales, the short-term economic activity of felling trees, generating capital and jobs in the immediate timeframe is a low-level micro scale that unfolds or takes place at a rapid rate, that is, over a short time period. The view of the forest in this scale is that of the individual tree that is seen as a commodity and turned into capital and profit. The view of the forest as a landscape/community ecosystem is one that unfolds over a much longer timeframe and at a much slower rate of development. It is, clearly, a different view from the economic one, and, following hierarchy theory, the longer term view and slower rated scale is understood to be a higher scale that is the context for the shorter timeframe, faster rated scale. In this scenario, this longer timeframe, slower rated scale corresponds to Norton's intermediate community scale. Following the proposed framework of scales noted above, the forestry industry, and the value set associated with it (economic prosperity, profit, jobs), is a local scale phenomenon.

On the other hand, another value set identified by the Committee in these recommendations is that related to sustainability, preserving habitats, biodiversity, watersheds, social uses of forests and protecting the forests for use by future generations. These values are all oriented to the long term and are associated with the communal scale. Note that I have begun to talk here about values. The phenomena that we are considering may be objects that can be assigned to a scale based on their rate characteristics as described earlier. For example, a tree, a forest and a mountain would correspond to local, communal and global scales, respectively. The phenomena may be actions or activities like cutting down of a tree or trees, the regeneration of a forest from seral to climax stage, and the slow weathering of a mountain range corresponding to local, communal, and global scales, respectively. In addition, the phenomena of concern may be the values ascribed, by some person(s), to these objects or actions. Using Roy et al.'s (1994) framework in this forestry issue we recognize a category 6 conflict - conflict over morality. This controversy revolves around the differing values held by various groups at loggerheads with one another. Having some way to place these values in some reasonable priority is necessary. As noted earlier, our values may also be viewed as scaled. Thus, we have values that are associated with short-term, long-term, and very long-term phenomena that could be categorized as local, communal, and global scale values. The conflict and controversy then is not just one of values but one involving the scales with which these values are associated. Thus, having some way to consider and prioritize the various scales is needed. This we have in the concepts of hierarchy theory as outlined in chapter 2, and this leads to the second question of the framework that aims to make explicit for normative use the scale that is contextual and those which are constituent.

This question asks that the relationship of the identified scales of concern be made explicit. Here, the moral agent is directed to be clear about which scale is constituent and which contextual, that is which scale is a part and which the whole. From hierarchy theory, it is understood that the communal scale is contextual for the local scale and the global scale is contextual for all. However, as we saw in the discussion about various current views on scales (chapter 2, third section), the broad ecological classification of three scales does not capture the richness of human social systems. We need to accommodate this in our framework, and being reflective about parts and wholes, constituents and contexts facilitates this accommodation. For example, in the case of forestry, the local scale activities of the forest industry unfold within the context of the forest ecosystems and natural landscapes, which are communal scale phenomena and so contextual for the local scale constituent. If we consider the values held by the forest industry, some could be considered social in nature, for instance, the protection and promotion of jobs for people. This, however, is a short-term concern that contrasts with the social values of protecting and promoting "cultural and spiritual" (p. 1) values and the notion of fairness toward future generations that are long-term concerns. The short-term activities of forestry unfold, as it were, within the context of the present society in general and as such are viewed as constituent to the context of the present society itself. Thus, forestry (and its social value set) can be characterized as a local-constituent scale and present society (and its value set, specifically the cultural, spiritual and concern for future generations) can be characterized as a local-contextual scale. The continuum of present to near and long-term future societies greys the boundary between local scale and communal scale, so it is reasonable to consider these longer term societal horizons of concern to rest in the communal scale. To distinguish them from the even more contextual ecosystemic aspects of the communal scale, the social aspects of this scale need to be identified as constituent, and therefore, following our framework, categorized as falling in the communal-constituent scale, while the broader natural context is categorized as a communal-contextual scale. Here we see the capability of conceptualizing scales, subscales in effect, within one ecological scale. Thus, we can identify a local-constituent scale, a local-contextual scale, a communal-constituent scale, and a communal-contextual scale.

The third question of the framework asks that the relationship between the contextual scale and its constituents be explicitly considered, specifically considering the nature of the activities that are required to not only protect, but also to promote each scale. We are seeking here to determine the effects of these activities, of supporting the values associated with each scale, and whether the integrity of the contextual scale is at risk. This approach brings in the prescriptive aspects of the concept of scale to be applied to the decision-making process. For example, in the Committee's report, the value set associated with the local-constituent scale is predominantly economic and consumer- oriented, with value given to profit, employment, regional development (seen as economic) and yearly addition to the "commercial forest base" (p. 1). This last activity of increasing the commercial forest base, that is, opening more forests to lumbering, is, of course, another way of referring to the act of cutting down trees. From the perspective of the ecosystemic communal scale (the context for the forest industry), increasing the commercial forest base is a definite challenge, at the very least full of uncertainty, to the overall integrity of the forest ecosystems involved.

Looking at the first two recommendations of the Standing Committee, I have identified three scales, a local-constituent scale (forest industry), a local-contextual scale (society), and a communal scale (ecosystem/landscape). It appears that the Committee supported the concept of undertaking forest management in Canada with emphasis on the communal scale and the values that are associated with it ("achieving sustainable forestry" p. xv). As a higher scale, the communal scale values would be considered preeminent over those of lower scales when it is evident that the integrity and good functioning of the higher scale, communal in this instance, is at risk or when uncertainty about this level of risk is great.

Recommendations 3 to 9

Although "the Canadian Constitution has assigned responsibility for forest management to the provinces" (p. 33), this cluster of recommendations recognized that the federal government still has a responsibility to the Canadian people regarding forest management. This is especially true in "the areas of industrial and regional development; international affairs including trade;

CASE STUDY: AN EXAMPLE OF SCALE   77

research and development (R&D); the environment; and the management of forest activity on federal lands and aboriginal reserves" (p. 33) as these relate to forest management in the country. Thus, recommendation 3 highlights the areas of specific concern for federal government involvement, and recommendations 4 to 9 present the specific actions for each of the areas of concern.

> Recommendation No. 3 states:
>
> That, notwithstanding the fact that jurisdiction over forest management lies with the provincial and territorial governments, the federal government through the Canadian Forest Service adopt a more proactive national leadership role in the forest sector to coordinate governments' efforts in such areas as R&D; effective domestic and international communications; public education; the development of national forestry data and sustainable development indicators; the collection of information on sustainable forestry program and biodiversity in other countries; and the negotiation of an international forest convention. (pp. xv and 34)

Looking at this recommendation, it is evident that all of the actions being suggested are immediate, short-term undertakings and as such are seen as low-scale activities that would be categorized to the local scale. The recognition of this as a constituent to the context of current social (local-contextual scale), long-term social (communal-constituent scale) and ecological (communal-contextual scale) values, as noted above for recommendations 1 and 2, identifies the communal scale as the context scale. As noted, within this communal scale the longer term social values would be characterized as constituent (communal-constituent scale) to the still broader ecological phenomena that we also value, for example forest sustainability and development, ecosystem and species diversity, and soil formation. These would be considered as a communal-contextual scale.

> Recommendation No. 4 stated:
>
> That the federal government fully support current efforts to obtain domestic and international certification for Canadian forest products through the Standards Council of Canada and the International Standards Organization. (p. 36)

This recommendation stems from the need to have in place some sort of "certification system" to ensure consumers of Canadian forest products that these products are "derived from forests that are sustainably managed" (p. 35). This assertion reflects the concerns in recommendation 3 for "effective domestic and international communications" and "public education." This action of establishing a certification system is assigned to the local scale. However, the concept of "forests that are sustainably managed" is one that is ascribed to the communal scale due to the implications of sustainability as discussed earlier. This recommendation, therefore, incorporates both local and communal scales. The communal scale, of course, is contextual. The development of certification systems and the process of public education do not appear to be activities that, in themselves, would challenge the integrity of the contextual scale.

Recommendations 5, 6, and 7 cluster together specifically around the issue of being able to identify and corroborate when sustainable forestry is achieved and educating the public in this regard. They are as follows:

> Recommendation No. 5: That the federal government work with the provincial and territorial governments to develop, on the basis of scientific indicators, the national data required to accurately measure and report on the achievement of sustainable forestry. (p. 37) ...
>
> Recommendation No. 6: That the federal government, in conjunction with the provinces and other stakeholders, design an independent review mechanism to assess the information contained in the National Forestry Data Base and the annual Report 'The State of Canada's Forest.'" (p. 38)
>
> Recommendation No. 7: That the federal government, in conjunction with the provinces/territories and other major stakeholders, launch an aggressive and comprehensive public education campaign to inform Canadians about the current state of Canadian forest management as well as the economic and environmental importance of a sustainable forest resource. (p. 38)

These three recommendations are all local-scale phenomena, are constituent in character and do not specifically challenge the communal or global scales. In fact, they may be supportive of these contextual scales by contributing to a better understanding of them with regard to what is required to maintain their integrity and good functioning.

Recommendations 8 and 9 further address the need for more knowledge and application of this knowledge to our forest management.

> Recommendation 8 states:
>
> That the R&D capabilities of the Canadian Forest Service be reoriented to place even greater emphasis on the sustainability of Canada's forest harvesting and regeneration practices, especially clearcutting. Special attention should be devoted to biodiversity; wildlife; and forest ecosystem and landscape management. (p. 39)

It is focused on the technology used in the lumber industry, specifically the consequences technology has on valued ecological phenomena. These ecological

CASE STUDY: AN EXAMPLE OF SCALE 79

values are located in the communal scale. Recommendation 8 therefore puts forward an action in the local scale (reorientation of R&D) in support of its contextual scale, the communal scale.

> Recommendation 9 states:
>
> That the federal government, the provincial governments, research institutes and the academic community work together to more effectively transfer ideas and technology to stakeholders active in the forest. The Model Forest network, among others, should serve as an appropriate two-way mechanism for such a transfer of knowledge. (p. 39)

The action focus of this recommendation is a local scale phenomenon as well. It is also supportive of the integrity of its contextual scale(s), both the social and ecological, of which it is a part with regard to advancement of knowledge. However, the simple transfer of ideas and technology to stakeholders active in the forest could be a challenge to the integrity of the ecological communal scale. Modern forestry technology with its emphasis on heavy equipment, and in Canada with an aggressive clear-cutting strategy, creates extensive damage to the soil, undergrowth, and dependent organisms, enough damage to challenge the ecological communal scale.

Recommendations 11 to 14

In their attempts to foster regional forestry development, the federal and provincial governments had, prior to 1994, established Forest Resource Development Agreements (FRDAs). These agreements were designed to "stimulate increased and sound forest management activity by various stakeholders" (p. 40) and the Committee notes in its report that many witnesses before it pointed out the "important benefits to the forestry sector" (p. 41) that the FRDAs provided. These benefits appear to have been, for the most part, economic as the Committee goes on to note that they "would like to see future funding [for the FRDAs] to be more targeted towards defined uses and to be conditional upon the demonstration of sustainable forestry practices" (p. 41). Thus:

> Recommendation No. 11:
>
> That the federal government negotiate a renewal of the Forest Resource Development Agreements (or equivalent arrangements) with the provinces for an additional five-year phase, and that the principal foci of the new programming be the development of forest ecosystem and landscape management techniques, and the continuation of financial assistance to private woodlot owners to encourage sounder forest management practices...

Recommendation No. 12:

> That in the future, the provision of assistance through reconstituted FRDA programming be conditional upon program recipients' demonstration of sustainable forestry practices. In the case of private woodlots, federal funding would be contingent upon the existence of an approved management plan. (p. 41)

Considering these two recommendations with the questions for scalar analysis in mind, it is evident that the activities referred to, the renewal of FRDAs and forestry, are occurring in the immediate, short-term micro scale. However, the committee has identified a set of values (sustainability and all it entails) that, at least in these recommendations, it believes should constrain these activities or establish limitations under which they take place. The sustainability value set that the committee is emphasizing here belongs to a broader, slower rated, long-term scale (communal scale) than the faster rated short-term scale (local scale) of forestry itself. The sustainability value set associated with the communal scale is contextual to the economic and technological value set of the constituent local scale.

Recommendations 13 and 14 deal with the recognition by the committee "that aboriginal peoples want greater involvement in decision-making regarding forest use" (p. 42).

> Recommendation No. 13: That the Minister of Natural Resources work together with other federal government departments and the Canadian Council of Forest Ministers to implement the aboriginal component of Canada's National Forest Strategy under strategic direction no. 7.

> Recommendation No. 14: That the national forest inventory be broadened to include an inventory of traditional land use activities. (p. 43)

Recommendation 13 is a statement of respect for aboriginal peoples, their needs, and values. As such, it recognizes aboriginal society and cultural values as a part of the whole of Canadian society, both of which, in scalar terms, are parts of a greater global human society. These unfold, that is, live out their lives, in terms of decades and intergenerational time and therefore from a scalar view are to be considered a higher order than the individual, short-term, economic scale. In this analysis, therefore, the value set respecting aboriginal society and culture is taken to reflect the intermediate, communal scale and therefore would be a context under which economic activities in the forest industry (local scale) take place. Similarly, Recommendation 14 can be interpreted as a communal scale. Here, the "traditional land use activities" being referred to, extend beyond "that of timber resource" (p. 42) (i.e., economic) to "the protection of the forest for future generations ... respect for biodiversity ... derived from native peoples' long-held views that land and forests should be viewed in a holistic way" (p. 41). That is to say, traditional land use activities express a value set that extends beyond the economic one associated with the local scale to a broader set of concerns associated with the communal scale. The communal scale, as I have delineated it, is the contextual whole within which lower order scales are active. The implications of the committee's recommendations here are that they at least have the perception that the integrity and good functioning of this contextual scale is at risk from the unconstrained activities of the constituent scale. It is well known that Canadian forestry practices have been aggressively criticized here and abroad, and this has hurt the forestry industry economically (short-term, micro scale). It is the committee's belief that such public pressure has succeeded in sensitizing governments and the domestic forest industry to the need to preserve environmental quality and the social values of the forest ... [and that] ... industry now realizes that forest management involves much more than simply sustaining timber yield, and that the forest must increasingly be viewed from a number of other 'value' vantage points such as biodiversity. (p. 2)

Recommendations 15, 16, and 17

To support the economic status of the Canadian forest industry, a local scale concern, the Committee believes that "it is absolutely vital that the positive message of Canadian forestry go out to overseas markets, so that the world can discover that Canada's forestry practices are as enlightened as anywhere else" (p. 45). The "positive message" here is an emphasis on sustainability of values beyond the micro economic timber values, that is, an emphasis on non-timber values which in scalar terms are intermediate communal scale concerns. The committee reports that Canada is "spearheading discussions on the development of internationally acceptable criteria for sustainable forestry" (p. 47) and states that what is needed is "an international agreement on the principles and standards of forest management that would be both transparent and scientifically defensible" (p. 46). From the committee's perspective, "such an International Convention on Sustainable Forestry would address the management, conservation and sustainable development of all types of forests" (p. 46). The committee also states that international agreements would need to specifically address topics such as:

- forestry practices, silviculture, and forest renewal;
- wilderness protection and biodiversity;
- protection of air, water, and soil;
- conversion of native forests to plantation forestry with exotic species;
- conversion of forest lands to agriculture and human settlement and
- monitoring compliance and enforcement. (pp. 46–47)

Recommendations 15, 16, and 17 are intended to capture this initiative and state:

> Recommendation No. 15: That the federal government, in conjunction with the provinces and territories, industry, environmentalists and other stakeholders, strive to consolidate the communications strategies currently employed in international markets into a single and effective campaign to promote Canada's forest management practices abroad.
>
> Recommendation No. 16: That as part of such a revamped international communications strategy, the federal government work in concert with Canadian forestry stakeholders and their international counterparts to organize a high-profile conference on sustainable forestry at the European Community level and/or within specific countries.... [and]
>
> Recommendation No. 17: That given the critical importance to Canada and the rest of the world of achieving an International Convention on Sustainable Forestry, the federal government increase its financial and personnel support of the Canadian negotiating team leading up to the conclusion of such an agreement. (pp. 46–48)

The activities (and the values that underlie them) elaborated on in the report, that is, those listed just above as topics needing to be specifically addressed, are a mix of both short-term local-scale concerns and intermediate long-term communal scale concerns. Forestry practices, silviculture, conversion of native forests to monocultures and to agriculture and settlement, and monitoring, compliance, and enforcement all take place in the immediate short term and are local-scale phenomena. On the other hand, forest renewal, wilderness protection, and biodiversity clearly unfold over a longer term and are considered intermediate communal-scale phenomena. Protection of water, air, and soil are at least communal-scale concerns and arguably could be considered global-scale concerns given the prolonged time frames in which these elements evolve and develop. It seems self-evident that the local-scale phenomena have at least the potential for risking the integrity and good functioning of the communal scale that is its context.

CASE STUDY: AN EXAMPLE OF SCALE 83

Recognizing Constraints

Table 4.1 summarizes the scales identified in the recommendations of the committee and their characteristics and whether or not the integrity of the greater whole is felt to be at risk based on criteria pertinent to the good functioning of the contextual scale, in this case these would be ecosystemic in nature. Hierarchy theory suggests that if the integrity of the contextual scale (whole) is at risk, seriously challenged, or clouded in uncertainty, then priority should be given to protecting this contextual scale. In the moral domain that we are considering, this is done by recognizing the values associated with the contextual scale as constraints to those activities of constituent scales that would undermine the integrity of its context. So a scalar view of an issue recognizes the requirements for maintaining the integrity of the contextual scale

Table 4.1 Summary of the scales identified in the recommendations of the committee

Recommendation	Scales	Character	Integrity of Whole
1, 2	Local-constituent	Constituent	Challenged
	Local-contextual	Context, constituent	
	Communal	Context	
3	Local-constituent	Constituent	Not challenged
	Local-contextual	Context, constituent	
	Communal-constituent	Context, constituent	
	Communal-contextual	Context	
4	Local	Constituent	Not challenged
	Communal	Context	
5, 6 and 7	Local	Constituent	Not challenged
8, 9	Local	Constituent	Not challenged, supported
11, 12, 13, and 14	Local	Constituent	Challenged
	Communal	Context	Supported
15, 16 and 17	Local	Constituent	Challenged
	Communal	Context, constituent	
	Global	Context	

as constraints on the activities of the local scale. That is, the value set of the local scale, as noted above, is sought within the limitations of the values that are associated with and support the integrity of the contextual scale.

The broadest scale, that described by Norton as the global scale, and here also referred to as the global scale, is the greater whole or context in which the previously described local and communal scales unfold. It is, therefore, necessary to look for effects on this broadest contextual scale of activities at the lower constituent scale.

Being explicit about the scales involved and comparing the values that are associated with them sets the stage for a reasoned prioritization of values based on the concept of constraints determined by the necessity of maintaining the integrity of the contextual scale. In this forestry case, the global and communal scales share the ecological values relating to climate, air, and water cycles and other ecological cycles, and the social and cultural values including a sustainable forest environment for the benefit of future generations. The local scale economic and social values relating to the business of forestry, profits, jobs, and development are not on common ground with those of its contextual scales. This circumstance, in part, explains the difficulties experienced for instance by the logging industry, loggers, and governments as they dialogue with environmentalists over the way our forests are treated. More specifically, if all the values held by these parties are recognized as valid and equal, how can decisions about forest management be made that supports them all? It is precisely here that the concept of scale is a useful addition to our current modes of ethical analysis. For example, Recommendations 1 and 2 demonstrate three scales: a local-constituent scale (the forest industry), a local-contextual scale (society), and a communal scale (ecosystem/landscape). The forest industry operates within the context of society and ecosystem scales and is therefore a part (constituent) and its actions would be constrained, as presented and reasoned in chapter 3, by the integrity needs of the contextual scales of which it is constituent. This dynamic would, for example, put limits on the forest industry with regard to indiscriminant logging when the values, integrity and good function of the contextual scales are in danger of collapsing. In other words, when, for instance, it is evident that logging, future jobs, and spiritual or aesthetic values (societal scale) are being lost, or when biodiversity and ecosystem life-support functions (ecosystem/landscape scale) are being lost. When this is the case, choices in favor of the values required to maintain the contextual scale(s) are necessary. When the contextual scales are not at risk, for instance, in a woodlot that is managed in a way that does not challenge the contextual scale's good functioning, then constituent scale activities occur following the value set of that local scale. The logging of the Christmas Mountains in New Brunswick referred to earlier is a case in point. Here, the forest industry has entered the last old growth forest in New Brunswick and by the value set of the industry, profit and jobs, the Christmas Mountains are being reduced to clear-cut. This situation severely alters the nature and function of the contextual scales. As a result, there will be little if any logging in the Christmas Mountains in the future and therefore employment security for future generations is also challenged. Similarly, human aesthetic and spiritual values in the Christmas Mountains are also diminished. These are local-contextual scale phenomena. The communal scale phenomena of ecosystem/landscape function and values like biodiversity, soil, and watershed integrity are similarly challenged by unfettered activity in the local-constituent scale (logging industry).

The descriptive nature of a scalar approach allows the explicit identification of different scales and the recognition of the values appended to the scales. The prescriptive nature of scale allows the recognition of all identified values as valid, yet provides reasons for prioritizing one value over another, contextual scale prior to constituent scale as presented in chapter 3. Based on this reasoning, in the example of the Christmas Mountains above, this would mean that there is a requirement to protect and promote the integrity and good functioning of the contextual scale. Here, this means that the communal (ecosystem/landscape) scale limits the activities of the constituent scales. For example, if the Christmas Mountains were made a provincial park dedicated to allowing the public to enjoy the recreational, aesthetic, and spiritual values of the forest (identified as local-constituent and local-contextual scale phenomena), these activities could carry on within the constraints of the requirements to protect and promote the ecosytem/landscape scale values of diversity, stability, and water-, air- and soil- cycles. Thus, as long as the forest is not overused, to the point of endangering the ecosystem function, people are free to enjoy it. If, on the other hand, such use is placing a stress on the integrity of the context (as has happened, for example, in the Grand Canyon), then limitations become active. We see this often as regulations on park use and activities allowed or prohibited. In the Grand Canyon, for example, there are now limitations on the number of visitors allowed in certain areas at certain times. If the Christmas Mountains are left open to logging, then protecting and promoting the integrity of the ecosystem/landscape will require specific limitations on the nature and extent of logging that takes place.

In brief, if the pursuit of the values of the constituent scale has no substantive ill effects on the integrity and good functioning of the contextual scale, then the values associated with the constituent scale can and ought to be supported and sought. Conversely, if the pursuit of the values of the constituent scale does, or is likely to have, substantive ill effects on the integrity and good functioning of the contextual scale, then the values of the contextual scale that maintain its integrity and good functioning ought to take priority and set constraints for the actions undertaken in the constituent scale.

From a scalar perspective, this means that the value set of the communal scale ought to have priority over those of the local scale and that those of the global scale ought to have priority over both of the constituent scales it incorporates when/if the integrity of the context is at risk. Since, in this forestry case, the values of the communal scale and global scale are coherent they can be taken as constraints on those of the local scale. This does not mean that the local scale values are invalid and should be ignored, only that they need to be tempered by the limits set by the values of its contextual (communal and global) scales. The image being sought, the moral theme being asked for, is a convergence of understanding of what constitutes best interest for all scales involved. This shared understanding may then facilitate the optimization of the integrity, good functioning, and natural potential of all scales, contextual whole, and constituent parts alike, similar to Sahtouris' concept of mutual consistency mentioned earlier. With regard to the committee's report, the effects of scale are evident. Contextual scales, as I have identified them, do receive attention; regulations and forest management practices to ensure long-term sustainability (Government of Canada, 1994, p. xi) are enacted as constraints on activities of, what I have identified as, the constituent scales. However, nowhere is it explicitly stated that contextual scale phenomena, like sustainability, biodiversity, and ecosystem life-support values, should be preeminent as the scalar approach I advocate would require. The committee seems to recognize what I have referred to as the descriptive aspect of scale, but appears not to recognize, give credence to, or even consider the possibility of what I have called the prescriptive aspect of scale. There is evidence that the committee identifies both short- and long-term phenomena related to forestry but sends mixed messages about the priority of these. The committee seems to struggle with its prioritizing of the values sets that it recognizes. For example, at one point the committee expresses the desire to protect the long-term communal scale represented by its discussion of sustainability, but then it seems to prioritize the short-term, local scale when it states that forestry practices should "satisfy" economic needs (local-constituent scale phenomena) while only "taking into account" social and environmental values (local-contextual and communal scale phenomena) (Government of Canada, 1994, p. 4). Following the reasoning outlined in chapter 3, the scalar approach suggested here would reverse the emphasis in this statement, that is, that social and environmental values be satisfied and economic needs be taken into account respecting the constraints established by satisfying the contextual scale phenomena.

The scalar approach fostered here, like all good problem solving and decision making approaches, encourages clear description of knowledge at hand and when information is missing and uncertainties exist, directs attention to the missing information and suggests appropriate research endeavors to fill the gaps in knowledge base. Recommendation 3, for instance, is in accord with the descriptive aspects of the scalar approach as discussed at the beginning of chapter 2. There is the convergence of the values of the short-term scale and the longer term scale, namely the value of knowledge and of obtaining as complete an information base as possible for decision-making. Furthermore, there is also the capability of identifying and characterizing different scales even within a single scale category by being explicit with the identification of parts and wholes. As described above, this approach facilitates a richer description of human society from a scaled perspective and opens the door for the use of this in a normative mode. For example, Recommendation 3 identifies several value phenomena, R&D, effective communications, public education, and the development of databases, among others. Using the framework of questions and the matrix that I suggest, it is evident that these phenomena are of a local nature (the first question), occurring here and now, and that they are constituent in character (the second question), part of broader contexts, the current societal one in which they unfold (a local-contextual scale) and even a broader long-term, intergenerational social context (a communal scale) and a broader yet ecological context (also a communal scale). Since the ecological scale is contextual for the long-term social scale, it is identified as a communal-contextual scale and the long-term social as a communal-constituent scale. This analysis lets us put these values in some order and with some rationale for doing so. Since carrying out this recommendation, that is acting on the local scale values, would not seem to have a deleterious effect on the higher contextual scales, and in fact may well be supportive of them, (the third question) there are no specific constraining factors from the contextual scales.

In the discussion leading up to Recommendation 4, the committee elaborated on testimony from Professors Baskerville and Weetman (p. 36) regarding forest industry certification. Baskerville and Weetman are reported to have

> called for the federal government to create an in-house Forest Management Institute which would have three functions: (a) certification of individual forests so as to ensure that these are being managed in a sustainable manner with respect to an array of values; (b) the development of certification procedures for non-timber values; and (c) periodic auditing of forest management activities. The principal objective would be to ensure that the forest can 'consistently deliver an array of values over a time horizon of the order of 100 years. (p. 36)

However, the committee favored "the industry-led process, on the basis of cost and the fact that the current drive to develop an international certification process under the auspices of the ISO is well underway" (p. 36). The committee recommended "that the federal government fully support current efforts to obtain domestic and international certification for Canadian forest products through the Standards Council of Canada and the International Standards Organization" (p. xv and 36). The Baskerville and Weetman suggestion is a certification that is meant to ensure the consistent delivery of "an array of values over a time horizon of the order of 100 years." They are imagining a moral theme that delivers, that is ensures, that values of many sorts, both timber (economic) and non-timber (ecologic, sociocultural and spiritual), from all scales are respected and achieved. This scenario is viewed on a time frame of 100 years and represents concern for and emphasis on the intermediate, longer term scale referred to as the community scale. Looking at these statements critically will uncover the scales involved and will help to resolve conflict or assign priorities because the prescriptive nature of this scalar approach directs priority to the contextual scale. For instance, take the values of economic profitability and biodiversity that are alluded to in this fourth recommendation. If we argue their merits as if scale did not exist, what would count as criteria to choose between them? Biodiversity has currency over a long term and is not easily recognized as valuable in the present short term in which the value of profit in hand can be appreciated. In economic terms, such long-term values are seen as externalities and are discounted in the short-term, so without considering the prescriptive guidance of scale, economic profitability is favored. If we take the scalar approach, we recognize that profitability is a local-constituent scale value and that biodiversity is a communal-contextual scale value that has preeminence. Profitability would therefore be limited by the necessity to protect and promote biodiversity.

The committee's recommendation, however, is for current activity in the present timeframe. This is chosen for economic reasons ("on the basis of cost") and the timeframe they consider appropriate relates to their concept of a "25-year plan" (p. 27). This focus is in the short-term time-frame of the individual, local scale. It is the practical and economic values of this scale that are given precedence. An impression that is supported by the committee's emphasis on the profound yearly addition to our commercial forest base and the importance this has to regional development and the country's economy, while at the same time describing the ecological importance of the forests and the need for sustainability (p. 1). Furthermore, in recommending "the industry-led process" there exists an open door for a conflict of interest that is left unrecognized or, at least, unacknowledged. Morally, conflict of interest is best dealt with by avoiding it and when that is not always possible, at least making it explicit and open to scrutiny. Scale is not active in this category of conflict of interest. One would hope that the international approach that the committee does recommend would help in this regard, but if their approach is also practical, economic and to a great degree "industry-led," then such openness to scrutiny might well be in question. In this fourth recommendation, we see a clear discrepancy between the values of the communal scale, as represented in Baskerville and Weetman's testimony, and those of the local scale. Furthermore, one can see in this example that supporting the local scale (constituent) is incommensurate with support for the communal scale (context). However, supporting the communal scale is not incommensurate with ongoing activity in the local scale. Thus, it is evident that if the moral theme our imaginations favor is the integrity and good functioning of all scales involved, then giving priority to the contextual scale as prescribed by the scalar approach suggested here facilitates this outcome. A scalar approach helps describe the nature of an issue and helps assign values priority.

Recommendations 5 through 9, like Recommendation 3, put forward the importance of gaining more information and knowledge about our forests, and how they might be managed in the present to achieve sustainability. Moreover, they champion support for public education about what we think we know. Furthermore, Recommendation 8 ("that the R&D capabilities of the Canadian Forest Service be reoriented to place even greater emphasis on the sustainability of Canada's forest harvesting and regeneration practices, especially clear-cutting. Special attention should be devoted to biodiversity; wildlife;

and forest ecosystem and landscape management" pp. xvi, 39) and to a lesser degree Recommendation 9 ("that the federal government, the provincial governments, research institutes and the academic community work together to more effectively transfer ideas and technology to stakeholders active in the forest ..." pp. xvi, 39) draw specific attention to values that are clearly long-term communal scale values that they believe should be more widely known. This concept seems to imply that there is some understanding that these longer term scales and the value sets associated with them should be more broadly incorporated into our decision making framework and database. In other words, as noted above, there seems to be a recognition of the descriptive aspect of scale, as I have referred to it, yet a failure to recognize the prescriptive aspect of scale that I am suggesting also exists.

Canada has developed a "Model Forest Program" (p. 39) the objective of which is to test and assess various "new approaches to sustainable development of the forest." (p. 39). This is an activity occurring now with the short-term objective to gain more knowledge and as such is not in conflict with the sustainability value set of the longer term, communal scale. From a scalar approach, the implication of the Model Forest Program and Recommendation 10 ("that to contribute to the achievement of sustainable forestry, the federal government enlarge the existing network of Model Forests, and ensure that at least one additional Model Forest be totally under the management of aboriginal people") is that sustainability is the key value being supported and striven for. This is not a value associated with the short-term, local scale. It is a complex value set incorporating all the ecologic, environmental values previously identified in the report (pp. 1–2) and is associated with the long-term, communal scale. From a scalar approach, therefore, the long-term, communal scale in which this value set is active is the context or whole in which the activity of forestry itself and the economic (local-constituent scale) and short-term social values it supports (local-contextual scale) occur. It is the value set of this communal scale that establishes constraints under which the local scale activities are free to act. Following hierarchy theory (see chapter 2), as long as the integrity of the contextual scale is not disrupted or at risk of disruption, the constituent scale is free to act on its own value set. In this situation, the economic values and goals of forestry are free to proceed but within the constraints set by the need to ensure sustainability.

In the hope of promoting regional forestry activity, provincial and federal governments have established Forest Resource Development Agreements (FRDAs). These are "designed to stimulate increased and sound forest management activity" (p. 40), and need to be renewed. The committee "would like to see future funding to be more targeted towards defined uses and to be conditional upon the demonstration of sustainable forestry practices" (p. 41). Similar to the discussion above regarding Recommendation 10, in terms of scale, the committee is suggesting here that the FRDAs ensure sustainable forestry practice by all stakeholders, implying an emphasis on the long-term scale (communal scale) over the short-term local scale. Identifying the scales in this way, that is applying the concept of scale, with its emphasis on explicit identification of contextual whole and constituent part(s), facilitates a reasonable prioritization of values, contextual over constituent, as outlined in chapter 3.

For example, the committee expressed the following conclusion in its report: "there is an urgent need to continue, in some instances accelerate, the development of sustainable forestry practices in a way that both satisfies economic needs and takes into account the many social and environmental values inherent in a well managed forest" (p. 4). This conclusion clearly sets out practices that are to be active in the here and now (local-constituent scale) to "satisfy" economic values, i.e., profit and jobs. The "social" values alluded to include tourism and recreation industries, hunting and trapping as well as cultural and spiritual values (p. 1). These are also, for the most part, local scale considerations. The short-term social values were earlier described as local-contextual scale phenomena. The cultural and spiritual values are more elusive and insofar as they can be attributed to the human spirit and species and seen as transgenerational may be considered in the broader communal scale, and were earlier considered a communal-constituent scale. The ecological values identified in the report include habitat for plant and animal species (including ourselves), clean air and water, stable soils, regulation of local climate, carbon sink function and biodiversity (p. 1). By my proposed framework of scales, these values clearly belong in the broader communal scale that is contextual for the local scale and, as more explicitly defined earlier are categorized as communal-contextual.

As noted, the committee concluded that the economic values, that is, local-constituent scale values, need to be "satisfied," whereas the ecological values, that is, communal-contextual scale values, are to be only "taken into account." The apparent conflict between scales evident in this analysis mirrors the on-going conflict between the forestry industry and environmentalists, between the political and corporate elite, and the interest groups of the silent majority. The phrase apparent conflict is used since, as discussed in chapter 3, a broad and convergent understanding of what constitutes the best interest of all scales removes this dynamic from a frame perspective of conflict over scale to one of conflict and controversy over best interest.

Not recognizing the various proposed scales complicates the conflict over values and frequently leaves the argument over values polarized and confrontational. The recognition of scale is not solely a descriptive endeavor. Accepting the notion of scale that I am fostering includes accepting the prescriptive nature of it as well, thereby providing guidance for resolving conflict and controversy. For any particular activity, explicitly asking the questions about what scales are of concern (What scale is the phenomenon of concern occurring in?) and what constitutes the whole and the parts (What is the context and what is the part?) facilitates assessing the relationship of these scales. Furthermore, once the explicit identification of the value-sets critical to the integrity and good functioning of each scale has been achieved, the values dialogue can incorporate the knowledge about the relationship between/among the scales.

From the view of scale being suggested here, the moral agent is challenged to recognize that the protection and promotion of the integrity and good functioning of the contextual scale places constraints on the constituent scales (Is the integrity of the contextual scale at risk?). This prescriptive nature of scale is applicable to the prioritization of values, and judgments about which are to be mandatory, which permissible, which tolerated, and which prohibited. It might be asked, who would accept this? It is evident, I think, that this line of argument is congruent with Hugo's view (see p. 22), with Callicott's correlative ethics (see p. 56), with Potter's (1971, 1988, 1990, 1999) emphasis on long-term survival, with Leopold's (1949/1970) exhortation to "think like a mountain," with Norton's (1989, 1990, 1992) policy advice for ecosystem management (see chapter 2), with Giampietro's (1994) use of hierarchy theory in sustainable development (see Chapter 3), with Rolston's (1988) view quoted on page 55 and with Donnelley's (1998) view quoted on page 112.

This view of scale is more than just being aware of a different or broader perspective and, therefore, is not the same as or subsumed by Roy et al.'s second category of conflicts in bioethics, based on partial perceptions (see chapter 1). The concept of scale presented here is based on hierarchy theory and directs priority to higher level scales that provide the context for and the constraints under which lower level scales exist. Action within the whole (contextual scale) proceeds freely insofar as it does not disrupt or destroy its own integrity (see Westra, 1994, for an extensive discussion of integrity as an ethical principle). If integrity is compromised it results in altered function of both contextual and constituent scales. As outlined in chapter 2, the altered function is expected to be at a lower level of complexity at all scales involved. Whether or not such reduction in complexity always occurs is really a moot point because it can never really be known until it has occurred, so it seems preferable, on prudence alone, to always behave as if reduced contextual scale complexity, stability and diversity are the expected. Therefore this whole/part scalar approach would normatively place priority on maintaining the integrity and good functioning of the contextual scale. This necessitates the identification of the aspects of this scale, including what would be considered its value-set, that are required for its integrity and good functioning and the positive and negative obligation to satisfy these conditions (protect and promote the value-set). Now, this of course, is not to be taken as an all or nothing approach, that is, as noted earlier, the ideal that this approach strives for is the protection and promotion of the integrity and good functioning of contextual and constituent scales alike. We might understand the contextual scale constraints as the edge of the envelope that the constituent scales are pushing against. Like the envelope, sudden (short-term), intense (overwhelming) pressure can break it apart, while gradual (long-term), persistent (but not overwhelming) pressure can expand it. The nature of the "mutual consistency" as insightfully suggested by Sahtouris (1989) strives for a balancing of the value-sets of the contextual scale (whole) and the constituent scale (part). This shared harmonization of values develops in such a way that the requirements for the maintenance of the integrity and good functioning of both are secured. This dynamic requires the moral agent to identify the specific values and needs for the integrity of each scale. The negotiation, a practical reasoning, incorporates the knowledge that the particular requirements for integrity of the contextual scale are constraints on the constituent scale which, otherwise, is free to act as required for its own good, using its own systems and expertise. In the moral domain, this means that the existing moral systems continue to be used as they now are, that the scalar approach does not replace current moral theory, and principled analysis but adds to it. From the committee's report I have identified several scales, as summarized in the table on page 83, and the practical reasoning encouraged by the scalar approach recommended here looks to the values of these scales. The dialogue involved aims to find the actions that protect and promote the integrity and good functioning of each scale given the limitations that each has to live with. The committee could have been more emphatic in stating the limitations on the forestry industry necessary to protect and promote the integrity of the ecosystem scale value of biodiversity and the intergenerational societal scale value of sustainability.

Daily and Norton (1994) offered an example of this approach in their paper entitled, "A value-based, multi scalar approach to forest management." They use hierarchy theory to model the scalar nature of values established by the public. Their proposal for a "scalar system of resource management" (p. 3) suggests that

> Local activities can be constrained by a hierarchy of longer-term, higher-level considerations. That is, in a hierarchical structure, management at the local level may vary considerably according to local desires as long as they do not interfere with objectives established for management at larger scales and over longer time frames. ... Basically, management at any particular scale focuses on sustaining valued material resources or features through time by ensuring, via scalar constraints, that management activities and exogenous impacts remain within the brunt of higher-level ecological dynamics operating in larger time frames. By recognizing ecological constraints, multi-scalar forest management can help to avoid the larger-scale impacts which result from exceeding a critical threshold at a lower level. (pp. 3–4)

Hierarchy theory's attention to scale, which is conceptually a question of parts and wholes, helps identify and define values associated with a contextual scale that may function as constraints on various constituent scales. Recognizing and accepting these values as morally significant enables normative guidance for living within these constraints.

Not considering scale as suggested here has resulted in The House of Commons Standing Committee on Natural Resources report (Government of Canada, 1994) not recognizing the scalar nature of the values it identifies. Also, not considering scale, the committee misses out on another mode of discernment between/among values and does not seem to appreciate the contextual nature of the communal (larger-longer ecological) scale. It, therefore, fails, despite some consideration of limitations (Recommendations 11 &12), to give any priority to the communal scale values that it does identify. From the scalar approach, the committee has implicitly recognized a difference between the value sets but is not explicitly aware of this difference as one of scale. The result of this situation is that the conflict between the value sets remains in the turmoil of finding a reasonable means of prioritizing the values, which may all be deemed of equal status. In being explicit about the scales involved, identifying the wholes and parts, and the fundamental values associated with each of them, we have such a means for reasonably placing one value or another as a priority and a constraint under which the other values are expressed. In so doing, we then can use our imagination to creatively establish a means of reaching a "mutual consistency" between the scales and the values associated with them.

Looking back at all these recommendations and the statements made by the committee, there is, from a scalar-analysis perspective, a clear concern for activities and values of the longer term communal scale. However, there is a clear sense of a mixed message about values priority. On the one hand, the committee appears to recognize and recommend emphasis on what I identify as a communal scale (a contextual scale). On the other hand, though, the committee proceeds to conclude that this scale should be taken "into account" while local scale actions and values (constituent scale) should be "satisfied" (p. 4). This may well have been stated this way for political reasons, but these political reasons are, themselves, local scale phenomena and, in the scalar analysis I am suggesting, do not take priority over higher scale phenomena. This emphasis on lower scale values despite a statement acknowledging higher scale values leaves the impression that, despite the recognition of the importance of the concepts of an ecosystem approach and sustainability, the committee does not really appreciate the implications of these concepts. There is the feeling that the lower scale economic and forest industry concerns (local-constituent scale) are most important. An understanding and appreciation for the concept of scale would help rectify this mixed message.

Taking the example noted above, scalar analysis would also argue that economic needs be satisfied, but would make it very clear that, since they are a low local-constituent scale concern, they must be satisfied within the constraints set by the values and integrity of the higher local-contextual, communal and global scales (the greater whole) of which they are a part. This scalar approach does not take the place of principled moral analysis within the local scale, but adds to it by emphasizing the extension of our moral concern to scales beyond the short-term and the overriding priority of these contextual scales.

Hierarchy theory and explicit attention to scale suggests that all the activities and values identified in the committee's report are legitimate. However, they exist in different scales and should not be argued in a cross-scalar (vertical coupling) fashion because the scales are not considered to be equal in their very nature. Higher scales are considered to hold values that set constraints on the values and activities of lower scales that are parts of the whole of the higher scale. In this way, attention to scale in our moral analysis aids in prioritizing values and activities by providing a reasonable means of deciding between and among values. An analysis of this sort would suggest that protection of water, air, and soil (a global scale concern) sets constraints on activities of the communal and local scales; and that protection of wilderness, biodiversity, sustainability and forest renewal (a communal scale concern) sets constraints on the activities involved in forestry practice and silviculture (a local scale concern).

In his analysis of sustainable development, Giampietro (1994) concludes that "what we need are new ways to elaborate better compromises among contrasting values, risks and aspirations that are emerging from explosive technological developments and the growth of humankind" (p. 625). He claims that "the high level of uncertainty faced when dealing with intergenerational analysis means that theoretical principles of ethics are of little help in providing practical solutions" (p. 622). Giampietro goes on to recommend "an enlargement of the purposes of scientific inquiry ... to include ... the different roles that different levels play in the hierarchy" (p. 625). The key to this, he believes, is "the perception of belonging to the hierarchy" (p. 623). It is my contention that the introduction of the concept of scale forces a critical reflection on the identification and meaning of parts and wholes; that is, on the nature of their integrity and good functioning and the value-sets that are integral to them. Recognition and acceptance of values beyond the human scale that constrain us in our scale of activity and particularly the priority to be given to contextual scale concerns expands our moral domain and our conception of what is in our own best interest. I suggest that such an approach is critical to Giampietro's process quoted above and is a necessary companion to perceiving that we belong to the hierarchy and to the broadened "scientific inquiry" that Giampietro, Norton (1992), and Daily and Norton (1994) stress. Furthermore, the concept of scale can provide new vigor to our moral reasoning within the narrow human spatiotemporal scale of present ethical systems because it allows us to identify scalar aspects or subscales within the human spatiotemporal scale. For example, we can see definite boundaries in rates at which life unfolds and in which parts and wholes are considered between individuals, families, communities, society, and species. We can consider the relationship of these varying scales and their associated values and principles. Having a sense or understanding of scalar systems, parts and wholes, fosters an ability to prioritize the scales and therefore also the values. This is particularly true not only in the field of environmental ethics as noted in this section on forestry, but also in medicine where distributing resources among individuals and groups is a perennial issue (conflict and controversy).

Discussion and Other Examples of Scale

Resource Allocation in Medicine

Health care and medicine immediately come to mind as endeavors in which the distribution of resources is an ongoing issue often involving substantial conflict. It is already conventional (Hiller, 1984) to view this topic in an hierarchical way, defining micro (individual), meso (institutional/community), and macro (societal) levels of consideration. Various ethical approaches to this issue have been elaborated with no clear means to choose among them. The arguments put forward stem from our present conception of ethical theory and principles, and the arguments span the subscales. For example, Outka's (1974/1976) argument on social justice focuses on principles of distributive justice at the individual, micro level. He did not provide an adequate discussion/recognition of other scales of consideration, as if the micro scale was the only one that mattered. Lemieux-Charles and Meslin (1993), while identifying micro, meso, and macro levels, focused on the relationship between the roles and obligations of caregivers. Jecker and Pearlman (1992) suggested a framework for rationing health care that, they say, is intended

> to clarify the context in which different rationing criteria apply. Proposals to ration high technology or ration non-basic care generally apply to legislators, hospital administrators, and government bodies who make choices about the distribution of scarce dollars between different health care categories. Proposals to ration based on medical benefit or the equality of persons generally apply to settings in which persons more directly involved with patient care and management distribute scarce health care resources between patients. (p. 91)

These are clear descriptions of the macro, meso, and micro scales of concern. In general, in most of the resource allocation debates, claims for macro allocation, emphasizing both societal health and optimal health for the greatest number given resources available, are confronted with claims from individual liberty and freedom to choose one's own actions, and these may stand against concepts of equality. There may be nothing other than an ideological leaning to choose between or among them. That is, if you are a libertarian then you are likely to view the issue with individual autonomy as your principal value and argue for free choice in regards to the health care resources that you may desire. The needs of another in relation to your specific needs and desires may carry little weight in your view. On the other hand, if you are more inclined to a communitarian viewpoint, then equality, equity, and fairness are likely to be your principal values. You would argue more for individual needs than for desires as determinants for resource allocation and some understanding that the community needs may dictate limits to what any one individual can expect.

The point I want to make here is that this is not just a conflict over certain values. Embedded in this, and I suggest preeminent in it, is the issue over scale itself.

In this resource allocation issue, the macro level, defined as the level of societal concern, is the contextual scale for both institutions and communities (meso) and individual (micro) scales. Following the framework proposed above, this macro scale could be categorized as a communal scale based on its long-term slow dynamics and its associated long-term socio-cultural and spiritual value sets. This is the scale in which "we protect, develop and nurture our sense of who we are as a culture ... and decide what kind of society we want to be" (Norton, 1994a, p. 15). The other two scales identified here, the micro and the meso, could be categorized as local-constituent and local-contextual, respectively. Within the context of the communal scale (macro scale in resource allocation parlance) the two local scales (micro and meso) unfold/behave following their own moral, economic codes but the communal scale has, in effect, a 'responsibility' to maintain its own integrity and good function. Therefore, certain behaviors by the subscales (subwholes or parts) that would threaten the contextual scale are constrained. We see these constraints as policy and legislation.

It is obvious here that the nature of the scales we are dealing with are substantially different from the scales addressed in the earlier chapter. There, I was considering scales from an ecologic point of view and the timeframes used to establish boundaries between scales were in terms of a few years to decades (local scale), decades to centuries (communal scale), and centuries to millennia (global scale). These may still be applicable scales to consider when the issue under deliberation is human interaction with the environment, but less likely to be applicable when dealing with an issue involving purely interhuman activity. In this scenario, the concept of scale is exactly that, a conceptual way of looking at the issue from a perspective somewhat different from what has been taken as routine in interhuman ethics. As demonstrated in the discussion on the forestry issue, it is possible, using the proposed framework for scale, to integrate the ecological and social aspects to accommodate the depth and richness of human social systems. Looking for the part-whole, constituent-context relationship whether in relation to ecological scales (local, communal, or global) or in relation to a single scale facilitates the development of this depth.

This conceptual, scalar approach does not eliminate or disregard present ethical systems or principled arguments. As a contextual, "correlative ethics" (Callicott, 1986, p. 411), it adds to them by insisting on scalar identification and clarity regarding context. Seeking to uncover the parts and the whole, the constituents and context, in each analysis potentially opens another door leading one to the deeper, fundamental, enduring questions regarding the nature of human nature: What are we? How did we get this way? What gives us meaning? What are we to become? What is the moral theme that ought to be followed here? These questions constitute human philosophical, theological, and spiritual thought. Insofar as the contextual scale provides the umbrella of meaning and relationship for the scales that compose it, it establishes the constraints on its subwholes (parts) through which it maintains its own integrity and good functioning. In essence, constituent scales (subscales, parts) have freedom of action to reach their optimum expression of integrity and good functioning, but within the limits of the constraints that exist by the requirements of the contextual scale to protect and promote its optimal expression of integrity and good functioning.

Reiterating previous points, the concept of constraints is integrally part of scalar considerations, just as it is in moral theory. Explicit attention to scale is explicit attention to the contextual whole and constituent parts and seeks to uncover the story of the relationships involved. It is therefore also historical. Our reality and our ethical systems are also historical narratives of sorts describing, explaining, and guiding interaction and relationships (Rolston, 1988). The facts of life from one scale may well become values at another scale. Just as in understanding the nature of light (wave and particle), one needs both concepts to grasp a picture of the emergent whole, explicit attention to scale helps in identifying and understanding the values of the identified scales, not as separate dichotomies but as moral complementarities in interdependent relationship to each other. I contend that the concept of scale may form a basis for convergence of contemporary moral theories (utility, obligation, liberal individualism, communitarianism, virtue, care, and so on) and may provide for the extension of human centered philosophies to include humanity's interactions with the natural world as well. A scalar view facilitates the understanding of each of these different moral schools as complementarities with a common ground in the notion that all need to contend with the issue of being both a part and a whole at one and the same time. The notion of parts and wholes could become a common thread for dialogue within and among the schools. Such dialogue is often the method for convergence of varying ideas to a new understanding. The focus of scale is on the identification of parts and wholes and the recognition that everybody and everything embodies both of these attributes at one and the same time and therefore must seek coherence between these scales.

For instance, if we look at our national health system as the contextual scale for the allocation of health resources to its subscales, communities, institutions, and individuals (local-contextual and local-constituent scales, respectively) the question regarding the contextual scale for our national scale needs to be explored. That is, the national system becomes recognized as constituent to some broader scale. In this case, the broader context of multinational and global human health and health care must establish its place within the context of even broader global biological systems and their integrity and good functioning. Practically, this concept implies the necessity of all of us taking the concerns of this broader scale to heart and mind and working out how its needs, as well as ours, at a lower scale, can be met. This need is exemplified by recent developments like the formation of the International Society for Ecosystem Health and Medicine, and conferences like Ecosystem Health and Medicine: Integrating Science, Policy and Management held in Ottawa, June 19–23, 1994. As well, one can now readily find articles and texts that suggest the need for seriously accounting for the effects on human health of environmental changes wrought by human activity (Garrett, 1994, Kain, 1995, Roy et al., 1994).

It is apparent that the issue of scale exists. Indeed, this is further supported by the identification and separation of the fields of Public Ethics, Social Ethics, and Clinical Ethics that Roy et al. (1994) made in their book Bioethics in Canada. Recognizing these as parts and wholes as previously described and using this to consider priorities would, in my view, help in the analysis and resolution of the cross-scalar (vertical coupling) conflicts with which we so often struggle. For example, let us return to the stories outlined in chapter 3 (pp. 48–49). The relationship between George Smith and the orthopedic surgeon Bob Green alluded to in the first two stories represents an issue in clinical ethics, a local, individual, and short-term scale, while the stories relating to Bob Green, the hospital Utilization Committee and the Board of Directors represent an issue in public ethics. Here, the concern is not at the bedside, but at a somewhat broader scale relating to the integrity and order of the functions of society (Roy et al., 1994), in this instance the institution of local orthopedic care and the overall institution of health care for the local population. The integrity and good functioning of this scale is being challenged by the consumption of the budget for total knee implants coming from the lower scale. The fifth story, in as much as it relates to the long-term nature of health care as a societal good, and potentially as it relates to global health, is one of social ethics where the concern is with the integrity and good functioning of society as a whole (Roy et al., 1994).

We can see the scalar approach at work already in the stories as they represent what is actually occurring in our health care system. Let us look at this from the position of the Board of Trustees that has received the petition from the Utilization committee and Dr. Green for more orthopedic prostheses. Using the three question framework, the Board asks the first question regarding the scale that is representative of the action being considered, and identifies this as a local scale action the critical values of which are the quality of life and comfort of patient Mr. Smith. This is important to the Board, and it would like to do what it can to further the realization of these values. In considering the second question of the framework, the Board takes note that this local scale concern is just one of many that are a part of a broader scale. That is, it is a constituent in a contextual scale that takes into its fold the whole institution. The mission and values of this include not only the needs and values of Mr. Smith and Dr. Green but those of other patients and caregivers, the reasonable use and distribution of its resources, and a caring attitude in dealing with its constituents. The Board further recognizes that, like its constituents, it too is a constituent of a yet broader provincial governmental contextual scale incorporating all the regional provincial health boards and other community and social concerns.

Question three asks about the integrity of each scale with particular concern for that of the contextual scales and as the Board considers the petition before it, the evidence they have is that the budget for prostheses is at its limit. Increasing the prosthesis budget will stress the institutional contextual scale excessively unless some other constituent scale can afford to give something up, or an increase in resources from its own contextual scale (government) is forthcoming. The loss of integrity of the institution represented by its inability to provide for the maintenance of its function would result in the loss of some, perhaps many, medical services to its constituents and an increase in their discomfort. The Board challenges itself, along with its constituents and its context, to try to work out some creative solution that will allow it to continue to function well and also meet the health needs of Mr. Smith and other constituents. Meanwhile, however, the budget restraints are limitations on desired actions of Mr. Smith that are handled by him and his doctor following established criteria for the priority of care.

Global Health

Earlier in this discussion, I commented on the issue of resource allocation in medicine, pointing out that the scalar approach of micro (individual), meso (institutional), and macro (societal, governmental) is already common parlance in this field. However, as pointed out above, from the scalar approach, this current model is incomplete. From the scalar approach recommended here, and the framework of questions suggested to delve into the issue, we are required to identify the context(s) for all scales, and especially the context of the scale in which the contemplated action is occurring. In the issue of human health and health care, the micro scale is currently understood to be that of the individual, which in the suggested framework is a local scale. The meso scale is that of the institution providing care and is also a local scale concern, but is contextual to the individual scale. Therefore, we can recognize the individual scale as a local-constituent scale and the institutional scale as a local-contextual scale. The macro scale is that of society as a whole. The spatial and temporal characteristics of the local-constituent scale are those of an individual human being and are quite narrow and unfold in terms of hours to a few decades, with most health decisions being made in an extremely short-term period of a few days to a few weeks. By our earlier discussion, this is considered a low level scale; one which you would expect to be part of some higher level scale that would characteristically unfold over a longer timeframe and whose concerns, integrity, and good functioning would be considered in relation to this longer timeframe. Currently, in health policy, the meso scale is institutional in nature. Insofar as the institution's activities and concerns may be considered to extend beyond the narrow timeframe of a single individual, and insofar as it is the institution that is the umbrella under which individuals seek and receive health care, it is contextual for the individual, micro scale. It is, therefore, categorized as local-contextual. However, it is not the whole. The broader scale of societal concern and activity is to be considered, too. This scale, currently referred to as the macro scale, is clearly a scale under which both the individual and institutional scales operate and carry on their function and development. This scale is different in magnitude from the local-contextual scale in that its size, area, and numbers are much larger and its temporal timeframe is longer insofar as we consider society and government to persist into the future and into intergenerational time. Following Norton's ecological categories of scales, this macro scale is really better understood as a middle level scale that is the context for all individual and institutional development. It is within this scale that we, as a body politic, "protect, develop and nurture our sense of who we are as a culture ... and decide what kind of society we want to be" (Norton, 1994a, p. 15). Still, within this same general middle-level scale is the broader global human community. Since this global level is far larger in area and number than regional, national, and state or provincial scales, it is appropriate to consider this at a higher level, within the meso ecological scale. Therefore, we can categorize the societal considerations as communal-constituent and the broader global human society as communal-contextual. However, is this where it ends? The scalar approach I am suggesting requires further inquiry into context. It is at this point that we can see the need to include the non-human systems that are part of the middle ecological scale in our analysis, since these involve not only broad areas spatially, but incorporate longer timeframes in their development, integrity, and good functioning. Furthermore, these systems are required for the integrity and good functioning of all life activities, including our own. These can be categorized as communal scales contextual even to the global society scale. As some of these ecological scales actually unfold over millennia and evolutionary-geological time, they become truly macro or global scales providing the context for all else and, thus, can be categorized as a global scale in the proposed framework.

All this may seem somewhat self-evident and somewhat trivial, but the point is that, from the scalar approach, each of these scales, even the ones that are in essence subscales of a recognized category, have requirements for the maintenance of their integrity and good functioning. These requirements include key values or sets of values associated with that scale and that are consistently striven for by each scale within the context in which it exists. Let us consider the complex issue of developing antibiotic resistance in bacteria. As a more concrete example of this, consider the case of the spread of methicillin-resistant *Staphylococcus aureus* bacteria.

In the spring of 1990, Jim Henson, the famed puppeteer and creator of the Muppets, died of a common bacterial infection caused by *Streptococcus* (a local scale phenomenon). This strain, however, was different than usual, as it possessed genetic resistance to the penicillin class of antibiotics to which *Streptococci* are usually susceptible. Furthermore, this genetic resistance appears to have come to *Streptococci* via another type of very common bacteria, *Staphylococci*. What we know is this: bacteria seem to be able to pass genetic information between them by simple cell wall contact during which a segment of DNA from one bacterium is passed to the other. This can occur not only between different strains of bacteria of the same family but also between different families. In 1952, pretty well 100% of *Staphylococci* were susceptible to penicillin antibiotics. However, by the late 1960s a large percentage was resistant to penicillin, and doctors started to use another antibiotic called methicillin. Nevertheless, by 1982 fewer than 10% of *Staphylococci* were susceptible. This tale is well told by Garrett (1994) who writes:

> Methicillin-resistant *Staphylococcus aureus* (MRSA) outbreaks increased in size and frequency worldwide throughout the 1980's. By 1990, MRSA would represent a clear economic and health crisis for hospitals all over the globe. The incidence of MRSA infections and deaths would soar steadily, spreading from massive urban medical centers outward, eventually reaching to suburban clinics and rural treatment

> centers. ... Significant MRSA problems were soon showing up in far-flung locations, from rural Ethiopia to Perth, Australia. By 1993 only one surefire *Staphylococcus* killer would remain: vancomycin. And even the reliability of vancomycin was in jeopardy, as some physicians reported the existence of MRSA strains that could not readily be cured with the last of the available anti-staph drugs. Switching from inexpensive penicillins to methicillin increased drug treatment costs for a typical patient approximately tenfold; changing to vancomycin meant turning to one of the most expensive antibiotics on the market. It was a burden in the wealthy countries, but not prohibitive. The increased cost was beyond the reach of poorer nations, however, rendering some staphylococcal infections, practically speaking, untreatable. (p. 412)

This vancomycin resistant *Staphylococcus* is now prevalent worldwide, even in my small hometown of Fredericton! This is now a problem that could be characterized as a local-contextual scale issue as it now relates to the integrity and good functioning of the social order, and in its worldwide distribution and potential intergenerational nature, could be considered a communal scale issue.

We also now know that what has contributed greatly to this problem has been the injudicious use of antibiotics. This has occurred not only in medicine, where prophylactic use has been widespread, but, arguably, perhaps even more damaging, in veterinary medicine and the meat production industry where they are used extensively, not to treat but to prevent illness and improve the market value of the meat produced. This has been a problem because it has increased the exposure of bacteria to antibiotics and established the exact conditions necessary for the development of strains resistant to antibiotics and the exchange of the genetic material that confers this resistance to other bacteria. Garrett observes:

> *Staphylococcus* wasn't the only bacterial organism that was successfully using plasmids, jumping genes, mobile DNA, mutations, and conjugative sharing of resistance factors to overcome whatever drugs *Homo sapiens* threw at them. In fact, by 1993 nearly every common pathogenic bacterial species had developed some degree of clinically significant drug resistance. And over two dozen of these emergent strains posed life-threatening crises to humanity, having outwitted most commonly available antibiotic treatments. ... It seemed that new strains of bacteria were emerging everywhere in the world by the late 1980's, and their rates of emergence accelerated every year. In the United States alone, such emergences were adding an estimated $200 million a year to medical bills because of the need to use ever more exotic—and expensive—antibiotics, and longer patient hospitalization for everything from strep throat to life-threatening bacterial pneumonia. When the costs of extended hospital care were added, the estimated increase due to antibiotic resistant organisms topped $30 billion annually. (p. 414)

This quite clearly is a disastrous problem for a patient and doctor at the bedside, that is, at a local-constituent scale. The scale of concern however is broader than this and relates to the integrity and good order of society's functional institutions, like health care as a common good, a local-contextual scale. Here, in Canada, with our relatively small population and socialized health care, patients with these types of infections get the best treatment and most efficacious drug, but what about our neighbors in less well developed countries, and future generations? In as much as this problem relates to global health and intergenerational concerns, it may be considered to be a communal scale issue as well. Quite clearly, this is also a disastrous problem for broader communities and nations, and is global in nature. It is clearly evident that this is a scalar issue, that is to say, it not only encompasses the individual, local-constituent scale, but also the institutional, local-contextual scale as well as the broader, societal communal-constituent scale that extends internationally and intergenerationally in space and time.

The use and misuse of antibiotics have drastic effects and since we are now well aware of these and see a need to change our behavior, the issue is largely an ethical one, not simply one of science and technology. However, at what scale is this issue to be addressed? Should individuals make the decisions about what antibiotics they should receive and when, or should this be left up to physicians, the medical community, the government, or drug companies? What values and ends are to be supported and which must be less so? Although we are concerned here with a problem that has developed over a very short number of years, it is one that affects potentially everyone on Earth as well as other species and is a potential determinant of our vision of humanity in the short term and intergenerationally too. That is to say, the contextual scale is at risk. It seems self-evident that this problem needs to be dealt with and decided upon based on the needs and values of the contextual scale (larger picture and timeframe). Argued from autonomy, a local-constituent value, there is no obvious way to intervene and halt the progress of antibiotic resistance among bacteria. People, it seems, will do what they deem in their best interest regardless of information that suggests they do otherwise. Similarly, a free market approach (another local-constituent value) will not help either, as Garrett (1994) describes (p. 440). Public health approaches (local-contextual scale values) too are not of benefit because this problem involves more than just medicine. The pharmaceutical industry, the agricultural industries, governments, and society are all part and parcel of this issue. In as much as government and society may be considered as stable, slowly developing backgrounds for the activities of the other participants in this issue, they can be considered contextual and categorized as communal scale phenomena. The concern in this issue is for the well-being of not only the local scale phenomena, but also the communal scale. Since the communal scale, too, is at risk, it is this scale's integrity and good functioning that needs to be supported and promoted. As presented in chapter 3, the overall integrity and good functioning of constituent scales is subsumed by the protection and promotion of the contextual scale. The challenge to all scales is to establish a mutual consistency that takes contextual scale constraints seriously while fostering peak development of constituent scales. In terms of human life and the example above this really boils down to the engagement of all relevant scales and parties in dialogue that aims to work out how each can realize its peak development, understanding the broadened notion of self-interest that the scalar approach requires. Recognizing that the constraints that are attendant with the maintenance of the integrity of the communal scale are prescriptive for activity at the local scale helps with the discernment among values by guiding the priority of values according to scalar integrity.

Maternal-Fetal Issues

Another example of how a scalar approach is helpful comes from my experience in clinical ethics rounds on the Labour and Delivery-Perinatal Unit. I routinely engage the staff on these units in dialogue about issues that come up in their daily duties and give them pause for concern and, frequently, much moral discomfort. One such issue is exemplified by the following case. This case is an amalgam of several real cases, and so names and some details have been changed to protect privacy and confidentiality. Anna is 24 years old and is pregnant. She is single, is not in a known relationship and has two other children at home, ages 2 and 4. Anna lives with her mother. It is not clear who is really looking after the children, but there is no clear evidence of neglect. Anna has been reticent in coming to the prenatal care clinic, and the nurses are very concerned about her and her fetuses, yes, Anna is going to have twins! One of the twins is marginal in size for gestational age, and the other is clearly suffering from intrauterine growth retardation, commonly referred to as IUGR. This condition often results in poor health post-partum and, if severe, can result in intrauterine death. The nurses caring for Anna note that even with the twin pregnancy she has only gained ten pounds. They recognize her extremely poor dietary habits and have tried their best to educate her in this regard and to arrange for all possible help, including social assistance. However, Anna continues to resist all their efforts stating, "No one is going to tell me how to run my life and what I need to do." The nurses have brought this case up at rounds because they feel a strong need to look out for the welfare of the fetuses (principle of beneficence), and feel completely helpless in doing so because of the need to respect Anna's right to self-determination (principle of autonomy). They have a very real fear that one of these fetuses is likely to die before delivery and the other is at very high risk of illness and disease. They see this whole scenario as a conflict, a confrontation, of maternal rights versus fetal rights, and cannot, at least emotionally, understand why the doctors or somebody can not "just deal with it and protect the babies." When they are encouraged to think about and imagine ways of considering this problem and perhaps resolving it, they do not find much solace in the concept of autonomy, and feel their attempts to seek the welfare of the fetuses and the mother are severely restricted. Doing everything they can still leaves them feeling they have not done enough for the fetuses. Intuitively, they cast blame, not only on the mother who seems to them to be uncaring for the "babies she is carrying" but more importantly for my argument, they blame "society." In terms of a principled approach to bioethics, the nurses frame the conflict as one between the autonomy of the mother and the right to life of the fetuses and their (the nurses') concern to protect and promote the well-being of these vulnerable human beings (beneficence). From Roy et al.'s (1994) framework of conflicts in bioethics, we can identify, here, a restricted value focus on the part of both the mother and the nurses. Autonomy seems to be the critical focus of the mother and the well-being and right to life of the fetuses from the standpoint of the nurses. This dynamic also reflects conflicts on the level of morality, that is, over which of these values/principles should have priority. In addition, we are aware of conflict on the level of ethos (worldviews) about "the status, destiny, and meaning of human life, or about the order of society and the purposes and roles of societal institutions in maintaining human community" (Roy et al., 1994, p. 34). For example, what is the status of the human fetus? Should each human fetus be protected? When and to what extent should the fetus be protected? And what role does society have in any of this?

Looking at this situation from a scalar view, the bedside issue is a local scale phenomenon and the values of the mother's autonomy and the welfare of the fetuses are local scale phenomena as well. Trying to decide which value/principle should be primary is a horizontal coupling exercise (intrascalar). There is no single objective means that clearly helps the nursing staff do this. Arguments for one or the other perspective come from belief systems, that is, from worldviews, as they are higher order concepts. As explained above, cultural and societal values and spiritual concerns are categorized, in scalar terms, as either local-contextual scale phenomena, if they are essentially short-term fast rated, that is they are able to adapt and change fairly quickly, or, as communal scale phenomena, if they are long-term, slow rated phenomena, that is, they change very slowly. Following the tenets explained earlier, these contextual scales are the ones that establish constraints for the activity/behavior in lower constituent scales. Thus, the activity/behavior of the local-constituent scale (mother, nurses) would be constrained by policy, for example, at the local-contextual scale (institutions, local/regional governments, professional institutions, and so on), while these would be constrained by the broader socio-cultural and legislative agreements of the communal level scale. This dynamic represents a means of choosing between values. Values of the constituent scale are active and are protected and promoted to the degree possible within the constraints set by the values of the contextual scale, often represented by policy and legislation.

In the case outlined here, there is an intuitive recognition by the nurses that this tension between autonomy and beneficence is one that is broader than the individual, here-and-now scale in which they work day to day. In my conversation with these nurses they seem to understand that this issue deals with our humanity, with our sense of who we are and what has meaning and is of value to us. Insofar as this sense transcends individuals and takes a long-term view, it is understood to represent a broader communal scale and it is only at this scale of decision that we can define and support the values that are integral to our understanding of human nature. The point that struck me as significant in this conversation with these nurses was their intuitive application of a scalar view. Without explicitly recognizing that such a boundary had been made, yet knowing that this was a bigger issue than the bedside, they could agree that it needed to be tackled at a broader level. It seems that we intuitively identify scale and that higher scales are for decisions that set policy, define limits, look to the long-term, and establish constraints as has been expressed throughout this book. We need, however, to make it an explicit endeavor, so that we can at the very least know at what scale particular issues need to be addressed, and therefore what particular value-sets are operative. The choice between and among these may still require intense dialogue, negotiation and balancing, but the application of the prescriptive aspect of scale as I elucidated earlier is likely to be important in understanding the relationships and priorities between and among these values. For instance, in this scenario, the local-constituent scale values have been identified as those connected with the principles of autonomy and beneficence. The communal scale (institution, society) values that might be identified as operative here are tolerance, compromise, and public health as they relate to the integrity and good functioning of the communal scale. Policy or legislation may be envisioned that emphasizes these values, setting the context under which autonomy and beneficence are to be understood.

Sterilization of Mentally Incompetent Patients

Another example of a scalar conflict in bioethics is the issue of sterilization of the mentally incompetent patient. The following case is not fictitious, but names and some details have been changed to protect privacy and confidentiality. Mr. and Mrs. Doe are the parents of Janey who is mentally challenged due to tuberus sclerosis, a genetic condition that she has inherited. Janey lives with her parents and it is expected that she will do so for as long as they are capable of caring for her because she is not capable of independent living. Janey, however, is quite sociable and likes to be out and about and with others including the opposite sex. Janey is in her late teens and her parents are concerned that she may be, or become, sexually active and become pregnant. They do not feel that Janey has the capability of understanding and appreciating the nature of pregnancy and parturition and certainly does not have the capability of caring for another human individual herself. Janey's family doctor agrees with this assessment and, as well, points out that there is about a fifty-percent chance that any child that Janey has will also inherit tuberus sclerosis and become mentally challenged. Caring for any children that Janey might have would almost assuredly fall on Mr. and Mrs. Doe. They do not feel they are capable, for many reasons, to carry that responsibility and fear that, if they had to, Janey's care would suffer. They feel that the integrity of the family unit itself would be at risk. They have explored the possibilities for birth control and feel that mechanical methods would be unreliable because they rely on Janey to implement; likewise, long-term oral birth control has potential health hazards and depends on Janey's cooperation, which may not always be forthcoming. Long-term birth control with implanted hormone capsules is a possibility but is moderately invasive and requires repeated implantations, a minor surgery requiring anaethesia, over the years, all of which Janey is likely to have difficulty in assimilating and may be overly traumatic to her. The constant supervision and chaperoning that would be necessary to be effective as birth control is considered to be tantamount to imprisonment and unduly restrictive and harmful to Janey's optimal development. Mr. and Mrs. Doe have approached Janey's doctor seeking permanent surgical sterilization, which they feel is the most appropriate method under the circumstances. Although it would entail an invasive surgical procedure, this is now done by minimally invasive techniques using small stab incisions and small scopes with minimal risks. Although this is likely to be traumatic to Janey, it is likely to be less so than late pregnancy and several hours of labour and/or a cesarean section.

Cases similar to this one have been heard at the level of the Supreme Courts in Canada and in Great Britain and are worth reviewing. This is well represented in Canada by the Case of Eve (E. [P.E.I.], 1979; Eve [P.E.I.], 1981; Eve v. Mrs. E., 1986) and an excellent review of this case is available in Roy, Williams and Dickens, *Bioethics in Canada* (1994). In this case the court was petitioned by Eve's mother to allow Eve to undergo permanent sterilization as she was potentially impregnable but not mentally capable of understanding and appreciating pregnancy, parturition, and child rearing. In short, the Supreme Court of Canada eventually ruled against this petition using, as justification, values and principles that are clearly within the individual micro scale (consent, right to reproduction, best interest). Best interest was considered very narrowly and no weight at all was given to the broader contextual scale and its values and how it impacted Eve's best interest, specifically the impact on and consequences to the family unit that cared for her and was her source of stability. The integrity, good functioning, and stability of this family unit, though not an absolute for her existence and support, are, at this time in her life, arguably the most critical values protecting and promoting her own development. A scalar analysis would require that an in-depth consideration of Eve as a whole and as a part of a larger contextual scale be undertaken. The moral significance of both levels is understood and appreciated by this attention to scale, and the proper course of action will be the one that supports the integrity and good functioning of all scales involved. The contextual scale must not be undermined, yet the lower level constituent scale (the individual in this case) must be free to strive for its optimum self-expression.

The Court, in this case, emphasized the constituent, individual, and personal values related to the right to reproduce, the privilege of pregnancy and parturition, and on consensual intervention despite the person's inability to comprehend and appreciate these. The Court, in short, could not conceive of the notion that non-consensual, non-therapeutic sterilization could ever be in a person's best interest, a narrow view of best interest from the scalar approach fostered in this book. When the contextual scale is deemed to be at risk of degrading change or collapse, or when the effects are quite uncertain, then its values should be considered a priority. Explicit attention to scale in the case of Eve would take a different view of the issue from the one of the Supreme Court. The notion of a broader understanding of best interest encouraged by a scalar approach could have provided support for non-therapeutic sterilization for Eve. However, the particulars of each case would determine the proper course of action. For example, the case of Janey Doe outlined above differs from Eve in that the patient also harbors a genetic illness with a fifty-percent chance of transmission and the strain on the contextual scale of caring for another disabled child would be greater. The ability of the context (family) to protect and promote Janey's well-being and development, including physical, emotional, and spiritual support and care could be seriously challenged.

A similar case has gone to the House of Lords in England (B [a minor], 1987), who, in their analysis, were quite critical of the Canadian Court's narrow interpretation of best interest and right to reproduce. The House of Lords dismissed the appeal emphasizing the importance of seeking the optimal development of, and quality of life, for B. What is missing from both of these legal cases is a consideration of the nature of the whole, the context, of which the respective individuals are a part. This consideration would be sure to discover the story that integrates the scales and explicates the nature and interdependency of the relationship between part and whole. The whole seeking to maintain its own integrity (so that it can continue to do what it does—in the case of Janey, for example, the family supporting and caring for Janey) while at the same time looking to optimize the development and the good functioning of its constituent part(s), that is, Janey's welfare.

In Janey's case, the context for her is her family. It is within this context that she lives, has her being, and finds meaning. The protection and promotion of Janey's optimal development is directly dependent upon the continued integrity of her family. She may find some meaning elsewhere, but a loss of family integrity surely would affect her life substantially and her overall care and optimal function would deteriorate. Such may well occur if Janey were to have a child that the family had to support. Preservation of the integrity of the family unit here, though not absolute, is a critical aspect of ongoing care for Janey. Sanctioning her sterilization may well be the appropriate thing to do if the contextual scale value of family integrity (the maintenance of Janey's context and thereby the promotion of her well-being) is recognized. Returning to the reasoning for supporting the priority of the contextual scale outlined in chapter 3, it seems evident that Janey is best served by a stable, caring and inclusive environment where support for her development and search for meaning can evolve as safely as possible.

By applying the three-question framework for eliciting the scales (local, communal, global), their character (contextual, constituent) and the state of their integrity we have a means of expanding the dialogue, giving more depth and breadth to our conception of best interest.

Regional Planning

An interesting article appeared in the November–December 1998 edition of the *Hastings Center Report*. It appeared as a special supplement entitled "Nature, Polis, Ethics: Chicago Regional Planning." The abstract to this paper states:

> As we plan for the futures of our metropolitan (and other) communities, we must come to understand more fully our moral and material relationships to the places we inhabit—the histories that have shaped both the people and place and what it will take to assure a future that can sustain "humans mutually flourishing in a mutually flourishing world". We must become "democratic ecological citizens" who think regionally about our interwoven, long-term responsibilities to humans and nature. (p. 1)

Here we see the exact elements of scale I have articulated here, that is, the need for an explicit understanding of the complex interdependencies and interrelationships that make up our individual, socio-cultural, and ecological biographies. The view of scale promoted here, and the analysis that it demands, leads to what I have called the contextual and constituent scales and their respective value sets and the constraints that are naturally related to the contextual scales within which the constituent scales have freedom of action. In the first of the articles in this special supplement, Strachan Donnelley (1998) referred to similar ideas, pointing out that "particular historical contexts and contingencies 'rule' evolution and life on earth ... that all human life and activity, perhaps without remainder, takes place *within* this reality and unique history" (p. s3, original emphasis). Donnelley further commented that:

> Regarding nature, most of us have yet fully to assimilate and appreciate the implications of evolutionary and ecological thought. ... We are still digesting all this and what it means for regional ecological and democratic citizenship. Yet this fledgling philosophical and ethical world view has had a decisive effect on our deliberations and explorations. Specifically, from a moral and civil perspective, we are taking the particularities of regional history—both natural and human—seriously. (p. s3)

Continuing with this line of thought, he observed:

> This historical "humans within nature" perspective is relatively novel territory in formal disciplinary scholarship, particularly ethics. It has animated our thinking and led us to

> break from the pack of more traditional philosophic and ethical approaches. We are groping for a new substantive "world view" ethics and a new art of "moral ecology," that is, the concurrent consideration and mutual coordination of long-term obligations to humans and nature. (p. s3)

I can observe only that these perspectives, in one form or another, are not new, as evidenced by the literature I have reviewed in this book, but that mainstream philosophy and ethics is finally hearing and responding to these voices. The concept of scale fits well within this perspective of humans within nature and is a useful way of making the consideration of context, constraints and freedoms an explicit aspect of our moral discourse in the field of bioethics. Explicit attention to scale, the deliberate characterization of parts and wholes, constituents and contexts, fosters the recognition of the particular contexts and contingencies that Donnelley acknowledges as reality.

· 5 ·

CONCLUSIONS, CRITIQUE, AND REFLECTION

Central Contention

This book begins with a personal story—that of looking over the landscape of a clear-cut mountain slope on the Queen Charlotte Islands in 1973. I saw the denuded slope eroding and sloughing into the ocean below and knew that something was wrong with this scene that a greater picture was being missed. This memory has been my steady companion over the years.

In the first chapter I took the stance that Potter has taken when he coined the term *bioethics*. That is, "that ethical values cannot be separated from biological facts" (Potter, 1988, p. 1), and I defined bioethics in terms of encompassing the moral dimensions of the biomedical, life, and environmental sciences. I provided support for this view from others and then went on to describe the seven categories of conflict suggested by Roy et al. (1994) as a framework for considering bioethical issues. Here I suggested the concept of "scale" might be considered a separate category as a primary source of conflict in bioethical issues. Several areas in which our present models for bioethical analysis and decision-making seem inadequate were highlighted and it was suggested that the common denominator in these was the failure to recognize the aspects of scale implicit in the issues under consideration. I closed the first chapter with an

Supporting Arguments

The Concept of Scale Helps in Considering Complex Systems

Chapter 2 considered more fully the nature of scale as I intend it to be understood and used. I explained that the concept of scale demands explicit description and clarification of the nature of relationships so as to define the parts and wholes, looking to identify a contextual scale. I pointed out that systems theory, and specifically hierarchy theory, from which I have extracted the concept of scale, is of great benefit in describing, investigating and managing complex living systems. This is to be expected in considering human systems and interrelationships with the natural world as well. Chapter 2 elaborated how scale has thus far been considered in various issues in bioethics. A description of "ecological" scales as elucidated by Norton and of "societal" scales as outlined by Hiller in the medical field was detailed. Callicott's view of ethics as correlative was also mentioned as, it too, is scalar in nature. I proceeded to suggest a framework of questions for considering scale that explicitly calls for the identification of parts and wholes, and their respective value sets. These notions of the rich moral flavor of a scalar analysis were expanded upon in chapter 3.

The Concept of Scale has Prescriptive Implications for Values Prioritization

The prescriptive implications of the concept of scale that I am proposing were developed in chapter 3 and the reasons to support these were outlined. That is, scalar analysis carries with it the notion that values integral to the integrity and good functioning of the contextual scale carry moral weight as constraints on the action or behavior of the constituent scales. This was supported by Norton's (1989; 1990; 1991; 1992; 1994a, 1994b; 1995) and Norton and Ulanowicz's (1992) work in ecological policy development. This is also evident in such a diverse field as archetypal psychology were it has been noted that "systems theory tells us that when any element of a system changes, the whole system has to reconfigure" (Pearson, 1998, p. 5). Further explanation, from a scalar approach, was given of the need to work through an issue with the understanding that a "whole" provides context and constraints for its parts yet, allows its parts to develop to their optimum state. Parts must strive to achieve their optimum development within constraints of the contextual scale or risk everything for every scale involved. The fact that any scale is at one and the same time both a whole and a part is morally significant and needs to be part of our deliberation in an explicit way. This is akin to Sahtouris's (1989) concept of striving to achieve and live within a "mutual consistency" (p. 27) whereby all components and wholes (constituents and contexts) work out a situation so that each optimizes its own integrity and good functioning without unduly compromising its own or that of the other(s). It is not at all clear that any of our present models of bioethical discourse, except Potter's plea to emphasize the long term, look at the issues this way. The point was made that our innate human ability for foresight carries both prudential and moral responsibility, and that explicit attention to scale fosters the use and optimal development of this capability. Furthermore, the explicit identification of scales and, particularly, contextual scales naturally fosters a long-term perspective that is consistent with the "global bioethics" for which Potter pleads.

Scales Are Evident in Bioethical Issues, and the Proposed Framework Helps to Elucidate Them

Chapter 4 presented an analysis of a government document dealing with forestry in Canada using the framework proposed for making attention to scale explicit. Here, the scalar aspects of the issue were made evident and their implications discussed. It was argued that the process of asking the moral agent to identify the parts and wholes and the relationship they share elucidates the scales involved. Critical reflection on these, bearing in mind the normative aspects of the concept of scale, has the potential for providing guidance in value prioritization. The argument was further supported in the remainder of chapter 4 with examples of issues in bioethics where this scalar approach could be used.

Adding Scale to Framework of Conflicts in Bioethical Issues

This book has been about the concept of scale and its implications for bioethics. It is apparent that the use of a simple framework to make scale an explicit component of bioethical analysis is feasible and worthwhile, particularly for the broad environmental-ecological issues where hierarchical systems and contextual scales are more readily evident. Certainly, in policy issues in the environmental-ecological realm, the use of scale in guiding inquiry and in helping to decide the question "at what scale should the decision be made?" appears to have substantial support. It is in this realm that it is indeed beneficial.

It seems, at first blush perhaps, somewhat less applicable in circumstances where the hierarchical system and contextual whole are less clear, particularly within human social systems where, for the large part, from an ecological perspective, most activity is occurring within the same scale. However, descriptive scalar approaches are used in medicine in resource allocation and implicit support for the whole is recognized. It is not much of a leap to be explicit about the scalar nature of the issue and integrate the prescriptive aspects of scale into the deliberations, particularly in the realm of values prioritization.

It is my conclusion that scale is indeed a real source of conflict in bioethical issues and that it should be considered as an eighth category in Roy et al.'s (1994) framework of types of conflicts in bioethical issues. Furthermore, the framework of questions advocated here for considering scale is an acceptable first attempt in integrating the concepts of scale into the analysis of various issues in bioethics. The explicit identification of context and constituents (wholes and parts) in a bioethical issue enables the nature of the relationships and interdependencies to be clarified and can provide normative assistance in decision-making about those issues. At the very least, this approach draws the decision-maker into explicit identification of the contextual scale and its value set and in so doing fosters our innate capacities of foresight and self-control. As such, I see this approach as an important addition to the general consideration of bioethical issues and a natural step toward the broader view of bioethics for which Potter pleads.

Critique and Reflection

Wilby (1994) presented "A Critique of Hierarchy Theory" as a systems theory, using the issue of trash disposal in Philadelphia as a case study. She defined seven tenets of hierarchy theory as a systems theory and critically appraises them. Some of her conclusions are pertinent to this reflection on attention to scale, as an additional tool in bioethical analysis, synthesis, and decision-making. Specifically, it is noted that the use of hierarchy theory is still in its infancy and its methodology is as yet not well defined, and in this regard the work identified in the literature outside the purely ecological sphere has been speculative. Wilby (1994) characterized hierarchy theory as perspectival in nature (as does Norton, 1994a) and suggests that this structure demands dialogue among perspectives to mitigate observer bias or at least to allow biases to be "openly and positively critiqued by other practitioners" (Wilby, 1994, p. 665). However, her analysis leads her to believe that hierarchy theory promotes and enhances self-reflection as well as the recognition of constraints and controls and the instabilities and disturbances that follow if these are transgressed. For the same reasons elaborated above, Wilby's conclusion that hierarchy theory promotes and enhances self-reflection applies equally to scale as I am suggesting its inclusion in bioethics.

Another problem is the semantics of the term *hierarchy*. This word has different connotations and meanings in different fields and often is seen in a negative light. Wilby identified this problem as well without giving any potential resolution for it. This difficulty over semantics can be reduced by the use of the term scale, as defined in this book.

It might be argued that a scalar perspective and consideration of parts and wholes is just Roy et al.'s (1994) second category of conflict in bioethical issues (see chapter 1), that is, a conflict based on partial perceptions. While certainly the concept of scale suggested here involves the "mistake of taking a part to be the whole" (Roy et al., 1994, p. 31), it is far more than simply "possessing limited information" (p. 31). The concept of scale, as I am using it, has its roots in ecological hierarchy theory. Not only does it address perspective (description), requiring an explicit identification and consideration of boundaries based on spatial extent and temporal rates of activity, but it is also prescriptive in nature, directing priority to the contextual (whole) scale when the integrity and good functioning of the whole is at risk. This dynamic is substantially different from the notion of partial perceptions and requires its own category of "conflict based on scale." Following Roy et al. (1994), a "practical observation about strategy in resolving conflicts" (p. 31) over scale is to

> *explicitly identify the scales involved in the issue, consider these from the standpoint of parts and wholes elucidating the constituent and contextual scale(s) and establish the fundamental values for each scale. Precedence is given to contextual scale values when it is evident that contextual scale integrity and good functioning is realistically at risk or uncertainty about its status is high.*

Many people coming from established philosophical/ethical theory will, no doubt, take issue with this concept of scale, particularly its prescriptive component, as I propose it. Constraints on individual freedom and choice are not taken lightly and will be seen as untenable. Therefore the whole concept of scale may be rejected forthright, but I have argued in chapter 3 that what is being called for is a broadened view of what constitutes best interest and that, given this view, constraints are not a rough-riding of individual liberty. Communitarians, on the other hand, are likely to find this approach weak. This is so because individual freedom is given free reign and is only constrained when evidence suggests the likelihood of a significant affront to the integrity of the whole (contextual scale). Alternatively, when uncertainty about integrity is high and prudence alone should dictate constraint and support of the contextual scale values until more certainty is established.

As outlined in chapter 3, there are several cogent reasons to support the notion that the context ought to have priority when it is at risk. The key point of understanding here is that this priority to context is not a heavy-handed, big stick taken to the constituents. Each aspect of the system, whole and parts, has a responsibility to the other as an extension of its own self-interest. Neither exists alone out of relationship and the integrity and good functioning of each are intimately connected. In "Nature" the balances are worked out by natural selection over many generations. In human social systems, such balancing of the integrity and good functioning of the contextual whole and constituent parts must be worked out through dialogue. This requires each aspect of the system to take the responsibility of caring for the integrity and good functioning of the other seriously. Although the distinctions of scale are not watertight, especially in social systems attention to scale encourages and enables each aspect of the system to take its responsibility for the other seriously.

It is my hope that these points are forceful enough to move those entrenched in either camp, libertarian and communitarian, or somewhere else. As noted earlier, in regard to the work of Strachan Donnelley, it may well be that these individuals and schools of thought have yet, as Donnelley (1998) says, "to fully assimilate and appreciate the implications of evolutionary and ecological thought" (p. s3). Explicit attention to scale, in my view, is a wholesome way of approaching bioethics precisely because it draws on the implications of evolutionary and ecological thought. Moreover, in so doing, it explicitly acknowledges how we are, at one and the same time, a part and a whole and it requires human moral agents to come to a full understanding and appreciation of such interrelationships. This approach results in "acknowledging human participation at many levels in the structure of reality" (Emberly, 1995, p. 20) and enables a responsible attempt to reach a state of "mutual consistency," as described by Sahtouris (1989, p. 27), in our life choices, as individuals and as regional and global societies. My approach to an ethical issue will clearly be different when I realize that I am the subject and the object, a part and a whole and not just one of these alone. My approach to an ethical issue will be different when I reflect critically upon the interrelationships that sustain me and give meaning to my life.

Several points of potential criticism of this approach have been considered here, but none of these is strong enough to dissuade me of the validity of a scalar approach to bioethical issues. Therefore, "Conflicts over Scale" should be added, as an eighth category, to Roy et al.'s (1994) framework of conflicts in bioethical issues. Furthermore, I suggest the following strategy for resolving such conflicts over scale:

Explicitly identify the scales (local, communal, global) involved in the issue; consider these from the standpoint of contextual whole and constituent parts; identify the fundamental values of each scale and give priority to the contextual scale when its integrity is challenged.

REFERENCES

Allen, T. F. H., & Starr, T. B. (1982). *Hierarchy theory: Perspectives for ecological complexity.* Chicago: University of Chicago Press.
Annas, G. J. (1995). Reframing the debate on health care reform by replacing our metaphors. *New England Journal of Medicine, 332,* 744–747.
Appelbaum, D., & Lawton, S. V. (Eds.). (1990). *Ethics and the professions.* Englewood Cliffs: Prentice Hall.
Arras, J. D. (1991). Getting down to cases: The revival of casuistry in bioethics. *Journal of Medicine and Philosophy, 16,* 31–33.
Auger, P. (1992). Hierarchically organized systems: The response to complexity. *Journal of Scientific & Industrial Research, 51,* 725–735.
B *(a minor)* (1987), 2 All England Reports 212.
Baier, A. (1985). *Postures of the mind.* Minneapolis: University of Minnesota Press.
Basson, M. D. (1979). Choosing among candidates for scarce medical resources. *The Journal of Medicine and Philosophy, 4,* 313–333.
Beauchamp, T. L., & Childress, J. F. (1994). *Principles of biomedical ethics* (4th ed.). New York: Oxford University Press.
Beauchamp, T. L., & Walters, L. (Eds.). (1994). *Contemporary issues in bioethics* (4th ed.). Belmont: Wadsworth Publishing Co.
Beecher, H. K. (1966). Ethics and clinical research. *New England Journal of Medicine, 274,* 1354–1360.
Blum, L. A. (1988). Gilligan and Kohlberg: Implications for moral theory. *Ethics, 98,* 472–491.
Bohm, D. (1980). *Wholeness and the implicate order.* London: Routledge & Kegan Paul.
Bohm, D. (1986a). The Implicate Order: A new approach to the nature of reality. In D.L. Schindler (Ed.), *Beyond mechanism: The universe in recent physics and catholic thought* (pp. 13–37). Lanham, MD: University Press of America.

Bohm, D. (1986b). Hidden variables and the Implicate Order. In D.L. Schindler (Ed.), *Beyond mechanism: The universe in recent physics and catholic thought* (pp. 144–156). Lanham, MD: University Press of America.

Bookchin, M. (1990). *The philosophy of social ecology: Essays on dialectical naturalism*. Montreal: Black Rose Books.

Bratton, S. P. (1992). Alternative models of ecosystem restoration. In R. Costanza, B.G. Norton, & B. D. Haskell (Eds.), *Ecosystem health: New goals for environmental management* (pp. 170–189). Washington, DC: Island Press.

Brook, A. (1987). Obligations to future generations: A case study. In W. Cragg (Ed.), *Contemporary moral issues* (2nd ed.) (pp. 446–458). Toronto: McGraw-Hill Ryerson. (Originally presented March 1980 at the Seminar on moral and ethical issues relating to nuclear energy generation, Canadian Nuclear Association [location not recorded]).

Brown, L. M., & Tappan, M. B. (1991). Commentary on Puka—justice debate. *Human development, 34*, 81–87.

Buber, M. (1970). *I and thou*. (W. Kaufman, Trans.). New York: Charles Scribner's Sons. (Original work published 1922)

Callahan, D. (1994, June 21). Nature as friend and nature as enemy: How do we choose? Paper presented at the First International Symposium on Ecosystem Health & Medicine, Ottawa, Canada.

Callicott, J. B. (1979). Elements of an environmental ethic: Moral considerability and the biotic community. *Environmental Ethics, 1*, 71–81.

Callicott, J. B. (1986). The search for an environmental ethic. In T. Regan (Ed.), *Matters of life and death: New introductory essays in moral philosophy* (2nd ed.) (pp. 381–424). New York: Random House.

Callicott, J. B. (1992). Aldo Leopold's metaphor. In R. Costanza, B. G. Norton & B. D. Haskell (Eds.), *Ecosystem health: New goals for environmental management* (pp. 42–56). Washington, DC: Island Press.

Canadian Medical Association. (1994). *Core and comprehensive health care services: A framework for decision making*. Ottawa: Canadian Medical Association Publishers.

Capra, F. (1982). *The turning point: Science, society, and the rising culture*. Toronto: Bantam Books.

Capra, F., Steindl-Rast, D., & Matus, T. (1991). *Belonging to the universe: Explorations on the frontiers of science and spirituality*. San Francisco: Harper.

Carbone, P. F. (Ed.). (1987). *Value theory and education*. Malabar: Krieger Publishing Company.

Carse, A. L. (1991). The "Voice of Care": Implications for bioethical education. *The Journal of Medicine and Philosophy, 16*, 5–28.

Charon, R., Banks, J. T., Connelly, J., Hawkins, A. H., Hunter, K. M., Jones, A. H., Montello, M., & Poirier, S. (1998). Literature and medicine: Contributions to clinical practice. In J.F. Monagle & D.C. Thomasma (Eds.), *Health care ethics: Critical issues for the 21st Century* (pp. 554–565). Gaithersburg: Aspen Publishers, Inc.

Cheney, J. (1989). Postmodern environmental ethics: ethics as bioregional narrative. *Environmental Ethics, 11*, 117–134.

Costanza, R., Wainger, L., Folke, C., & Maler, K. (1993). Modeling complex ecological economic systems: Toward an evolutionary, dynamic understanding of people and nature. *BioScience, 43*, 545–555.

Coufal, J. E., & Spuches, C. M. (1995). Ethics in the forestry curriculum: A challenge for all foresters. *Journal of Forestry, 93* (9), 30–35.

Cragg, W. (Ed.). (1987). *Contemporary moral issues* (2nd ed.). Toronto: McGraw- Hill Ryerson.

Crigger, B. J. (1996). Where do moral decisions come from? *Hastings Center Report, 26* (1), 33–38.

Dailey, R. G., & Norton, B. G. (1994). A value-based, multi-scalar approach to forest management. Unpublished paper, School of Public Policy, Georgia Institute of Technology, Georgia.

Devall, B., & Sessions, G. (1985). *Deep ecology.* Salt Lake City: Gibbs Smith.

Dienhart, J. W. (1982). *Cognitive approach to the ethics of counseling psychology.* Washington, DC: University Press of America.

Donnelley, S. (1998). Civic responsibility and the future of the Chicago region. Hastings Center Report, 28 (6), Special Supplement S2-S5.

Dumanski, J., Pettapiece, W. W., Acton, D. F., & Claude, P. P. (1993). Application of agroecological concepts and hierarchy theory in the design of databases for spatial and temporal characterisation of land and soil. *Geoderma, 60,* 343–358.

E. (P.E.I.) (1979), 10 Reports of Family Law (2d) 317.

Ehrenfeld, D. (1992). Ecosystem health and ecological theories. In R. Costanza, B.G. Norton, & B.D. Haskell (Eds.), *Ecosystem health: New goals for environmental management* (pp. 135–143). Washington, DC: Island Press.

Ehrenfeld, D. (1994, June 21). The marriage of ecology and medicine: Will they be compatible? Paper presented at First International Symposium on Ecosystem Health and Medicine, Ottawa, Canada.

Eisenberg, E. (1998). *The ecology of Eden.* Toronto: Random House.

Emberley, P. C. (1995). *Values education and technology: The ideology of dispossession.* Toronto: University of Toronto Press.

Englehardt, H. T. (1988). Foreword to Global Bioethics, by V.R. Potter. Lansing: Michigan State University Press.

Eve (P.E.I.) (1981), 115 Dominion Law Reports (3d) 283.

Eve v. Mrs. E. (1986), 2 Supreme Court Reports 5.

Faber, M., Manstatten, R., & Proops, J. (1992). Toward an open future: Ignorance, novelty, and evolution. In R. Costanza, R. G. Norton, & B. D. Haskell (Eds.), *Ecosystem health: New goals for environmental management* (pp. 72–96). Washington, DC: Island Press.

Fischer, F. (1990). *Technocracy and the politics of expertise.* Newbury Park: Sage Publications.

Fletcher, J. (1966). *Situation ethics: The new morality.* Philadelphia: The Westminster Press.

Fox, J. (1992). The Problem of scale in community resource management. *Environmental Management, 16,* 289–297.

Frye, N. (1987). Canada: New world without revolution. In W. Cragg (Ed.), *Contemporary moral issues* (2nd Ed.) (pp. 404–408). Toronto: McGraw-Hill Ryerson Limited. (Reprinted from *Division on a ground: Essays on Canadian culture,* 1982. Toronto: House of Anansi Press).

Gardiner, R.W. (1990). Between two worlds: Humans in nature and culture. *Environmental Ethics, 12,* 339–352.

Gardner, H. (1995). *Leading minds: An anatomy of leadership.* New York: Basic Books.

Garrett, L. (1994). *The coming plague: Newly emerging diseases in a world out of balance.* New York: Farrar, Straus and Giroux.

Gauthier, D. (1985). Maximization constrained. The rationality of cooperation. In R. Campbell & L. Sowden (Eds.), *Paradoxes of rationality and cooperation.* (pp. 73–93). Vancouver: University of British Columbia Press.

Giampietro, M. (1994). Using Hierarchy Theory to explore the concept of sustainable development. *Futures, 26,* 616–625.

Gibbs, J. C. (1991). Toward an interpretation of Kohlberg's and Hoffman's moral development theories. *Human Development, 34,* 11–104.

Gilligan, C. (1982). *In a different voice: Psychological theory and women's development.* Cambridge: Harvard University Press.

Gillis, A. M. (1990). The new forestry. *BioScience, 40,* 558–562.

Glaser, J. W. (1995). Phase II of bioethics: The turn to the social nature of individuals. *Bioethics Forum, 11*(3), 12–21.

Goodin, R. E. (1987). Ethical principles for environmental protection. In W. Cragg (Ed.), *Contemporary moral issues* (2nd ed.) (pp. 409–421). Toronto: McGraw-Hill Ryerson. (Condensed version of No moral nukes. Ethics, 1980, 90, 417–449).

Gorman, M. L. (1989). *Forestry and the ethics of coexistence: The Nova Scotia experience.* Master's thesis, Faculty of Environmental Studies, York University, Toronto.

Gorovitz, S. (1977). Bioethics and social responsibility. *The Monist, 60,* 3–15.

Government of Canada. (1994). *Canada: A model forest nation in the making.* Report of the House of Commons' Standing Committee on Natural Resources. Ottawa: Queen's Printer for Canada.

Green, R. M. (1977). Intergenerational distributive justice and environmental responsibility. *BioScience, 27,* 260–265.

Haigh, M. J. (1987). The holon: Hierarchy Theory and landscape research. *Catena Supplement, 10,* 181–192.

Hannon, B. (1992). Measures of economic and ecological health. In R. Costanza, R. G. Norton, & B. D. Haskell (Eds.), *Ecosystem health: New goals for environmental management* (pp. 207–222). Washington, DC: Island Press.

Hardin, G. (1968). The tragedy of the commons. *Science, 162,* 1243–48.

Harrington, R. F. (1990). *To heal the earth: The case for an earth ethic.* Surrey: Hancock House Publishers.

Haskell, B. D., Norton, B. G., & Costanza, R. (1992). Introduction: What is ecosystem health and why should we worry about it? In R. Costanza, B. G. Norton, & B. D. Haskell (Eds.), *Ecosystem Health: New goals for environmental management* (pp. 3–20). Washington, DC: Island Press.

Hiller, M. D. (1984). Ethics and health care administration: Issues in education and practice. *Journal of Health Administration Education, 2* (2), 148–192.

Hoffmaster, B. (1990). Morality and the social sciences. In G. Weisz (Ed.), *Social science perspectives on medical ethics* (pp. 241–260). Netherlands: Kluwer Academic Publisher.

Hogan, R. (1975). Moral development and the structure of personality. In D. J. DePalma & J. M. Foley (Ed.), *Moral development: current theory and research* (pp. 153–167). Hillsdale: Laurence Erlbaum Associates.

Holling, C. S. (1992). Cross-scale morphology, geometry, and dynamics of ecosystems. *Ecological Monographs, 62*(4), 447–502.

Hugo, V. (1988). Epigraph to *Environmental Ethics,* by Holmes Rolson, III. Philadelphia: Temple University Press. (Original work published 1890)

Jecker, N. S., & Pearlman, R. A. (1992). An ethical framework for rationing health care. *The Journal of Medicine and Philosophy, 17,* 79–96.

Jonsen, A. R., Siegler, M. & Winslade, W. (1992). *Clinical ethics: A practical approach to ethical decisions in clinical medicine.* New York: McGraw-Hill.

Jonsen, A. R., & Toulmin, S. (1988). *The abuse of casuistry: A history of moral reasoning.* Berkeley: University of California Press.

Kafatos, M., & Nadeau, R. (1991). *The conscious universe: Part and whole in modern physical theory.* New York: Springer-Verlag.

Kain, K. C. (1995). Emerging pathogens: The birth of plagues. *Annals of the Royal College of Physicians and Surgeons of Canada, 28* (3), 141–145.

REFERENCES

Karr, J. R. (1992). Ecological integrity: protecting Earth's life. In R. Costanza, B. G. Norton, & B. D. Haskell (Eds.), *Ecosystem health: New goals for environmental management* (pp. 223–238). Washington, DC: Island Press.

Katz, J. (1994). Reflections on unethical experiments and the beginnings of bioethics in the United States. *Kennedy Institute of Ethics Journal*, 4(2), 85–92.

Kaufman, W. (1970). *I and Thou, Martin Buber: A new translation with a prologue and notes.* New York: Charles Scribner's Sons.

Klijn. F., & de Haes, H. (1994). A hierarchical approach to ecosystems and its implications for ecological land classification. *Landscape Ecology*, 9 (2), 89–104.

Kluge, E-H. W. (1992). *Biomedical ethics in a Canadian context.* Scarborough: Prentice-Hall.

Koestler, A. (1967). *The ghost in the machine.* New York: MacMillan.

Koestler, A. (1978). *Janus.* London: Hutchinson.

Kohlberg, L. (1981). *The philosophy of moral development.* San Fransisco: Harper and Row.

Kuhn, T. S. (1970). *The structure of scientific revolutions* (2nd ed.). Chicago: The University of Chicago Press.

Last, J. (1993). Global change: A challenge for the medical profession. *Annals of the Royal College of Physicians and Surgeons of Canada*, 26 (1), 13–16.

Lazlo, E. (1972). *The systems view of the world.* New York: G. Braziller Inc.

Lemieux-Charles, L., & Meslin, E. M. (1993). Ethical issues faced by clinicians/managers in resource allocation decisions. *Journal of Hospital and Health Services Administration*, 38(2), 265–285.

Leopold, A. (1970). *A Sand County almanac.* New York: Ballantine Books. (Original work published 1949)

Levine, C. (1993). *Taking sides: Clashing views on controversial bioethical issues.* (5th ed.). Sluice Dock: The Dushkin Publishing Group.

Lilienfeld, R. (1978). *The rise of systems theory.* New York: John Wiley.

Loevinsohn, M. E. (1994). Climatic warming and increased malaria incidence in Rwanda. *Lancet*, 343, 714–718.

Manga, P. (1987). Medicare: ethics versus economics. *Canadian Medical Association Journal*, 136, 113–116.

MacIntyre, A. (1984). *After virtue.* (2nd ed.). Notre Dame: University of Notre Dame Press.

Mappes, T., & Zembaty. J. (1992). *Social ethics: Morality and social policy* (4th ed.). New York: McGraw-Hill.

Marietta, D. E. (1979). The interrelationship of ecological science and environmental ethics. *Environmental Ethics*, 1, 195–207.

McMichael, A. J. (1993). *Planetary overload: Global environmental change and the health of the human species.* New York: Cambridge University Press.

Midgley, M. (1987). Duties concerning islands. In W. Cragg (Ed.), *Contemporary moral issues* (2nd ed.) (pp. 434–446). Toronto: McGraw-Hill Ryerson. [Reprinted from R. Elliott & A. Gare (Eds.). *Environmental philosophy*, 1983. University Park, PA: Pennsylvania State University Press.

Milbrath, L. W. (1989). *Envisioning a sustainable society.* Albany: State University of New York Press.

Miller, A. (1991). *Personality types. A modern synthesis.* Calgary: University of Calgary Press.

Miller, A. (1999). *Environmental problem solving: Psychosocial barriers to adaptive change.* New York: Springer.

Narveson, J. (1987). Moral philosophy—what it is and why it matters! In W. Cragg (Ed.), *Contemporary Moral Issues* (2nd ed.) (pp. 422–433). Toronto: McGraw-Hill Ryerson.

Nicolis, G., & Prigogine, I. (1977). *Self-organization in non-equilibrium systems: From dissipative structure to order through fluctuations.* New York: Wiley.

Norton, B. G. (1989). Intergenerational equity and environmental decisions. A model using Rawls' Veil of Ignorance. *Ecological Economics, 1,* 137–159.

Norton, B. G. (1990). Context and hierarchy in Aldo Leopold's theory of environmental management. *Ecological Economics, 2,* 119–127.

Norton, B. G. (1991). *Toward unity among environmentalists.* New York: Oxford University Press.

Norton, B. G. (1992). A new paradigm for environmental management. In R. Costanza, B.G. Norton, & B.D. Haskell (Eds.), *Ecosystem health: New goals for environmental management* (pp. 23–41). Washington, DC: Island Press.

Norton, B. G. (1994a, April 19). A scalar approach to ecological constraints. Paper presented at the Workshop on Engineering Within Ecological Constraints, National Academy of Engineering, Washington, DC.

Norton, B. G. (1994b, June 22). Ecosystem health and social values: at what scale?" Paper presented at the First International Symposium on Ecosystem Health & Medicine, Ottawa, Canada.

Norton, B. G. (1995). Evaluating ecosystem states: Two competing paradigms. *Ecological Economics, 14,* 113–127.

Norton, B. G., & Ulanowicz, R. E. (1992). Scale and biodiversity policy: A hierarchical approach. *Ambio, 21,* 244–249.

Odum, E. P. (1989). *Ecology and our endangered life-support systems.* Sunderland, MA: Sinauer Associates, Inc.

O'Neill, R. V., DeAngelis, D. L., Waide, J. B., & Allen, T. F. H. (1986). *A hierarchical concept of ecosystems.* Princeton: Princeton University Press.

Outka, G. (1976). Social justice and equal access to health care. In T. A. Shannon (Ed.) *Bioethics* (pp. 373–395). New York: Paulist Press. (Reprinted from *The Journal of Religious Ethics,* 1974, 2(1), 11–32).

Page, T. (1992). Environmental existentialism. In R. Costanza, B. G. Norton, & B. D. Haskell (Eds.), *Ecosystem health: New goals for environmental management* (pp. 97–123). Washington, DC: Island Press.

Pattee, H. H. (Ed.). (1973). *Hierarchy theory: The challenge of complex systems.* New York: G. Braziller.

Pearson, C. S. (1998). *The hero within: Six archetypes we live by* (3rd ed.). New York: Harper San Francisco.

Pellegrino, E. D. (1985). The virtuous physician and the ethics of medicine. In E. E. Shelp (Ed.), *Virtue and medicine: Exploration in the character of medicine. Philosophy and Medicine Series, No. 17.* (pp. 243–255). Boston: D. Reidel Publishing Co.

Potter, V. R. (1971). *Bioethics: Bridge to the future.* Englewood Cliffs: Prentice-Hall.

Potter, V. R. (1975). Humility with responsibility—A bioethic for oncologists: Presidential Address. Cancer Research, 35, 2297–2306.

Potter, V. R. (1988). *Global bioethics. Building on the Leopold legacy.* East Lansing: Michigan State University Press.

Potter, V. R. (1990). Getting to the Year 3000: Can global bioethics overcome evolution's fatal flaw? *Perspectives in Biology and Medicine,* 34, 89–98.

Potter, V. R. (1999). Bioethics, biology, and the biosphere. *Hasting Center Report,* 29(1), 38–40.

Puka, W. (1991). Interpretive experiments: probing the care-justice debate in moral development. *Human development,* 34, 61–80.

Randall, W. L. (1995). *The stories we are: An essay on self-creation*. Toronto: University of Toronto Press.
Raphael, D. D. (1981). *Moral philosophy*. Oxford: Oxford University Press.
Rawls, J. (1972). *A theory of justice*. Oxford: Oxford University Press.
Regan, T. (Ed.) (1986). *Matters of life and death: New introductory essays in moral philosophy* (2nd ed.). New York: Random House.
Reich, W. T. (1994). The word " bioethics": Its birth and the legacies of those who shaped its meaning. *Kennedy Institute of Ethics Journal, 4*, 319–336.
Reich, W. T. (Ed.) (1995). *The encyclopedia of bioethics*. New York: Simon & Schuster MacMillan.
Rich, J. M., & DeVitis, J. L. (1985). *Theories of moral development*. Springfield: Charles C. Thomas.
Rolston, H. (1988). *Environmental ethics*. Philadelphia: Temple University Press.
Roy, D. J., Williams, J. R., & Dickens, B. M. (1994). *Bioethics in Canada*. Scarborough: Prentice Hall Canada.
Sagoff, M. (1992). Has nature a good of its own? In R. Costanza, B. G. Norton, & B. D. Haskell (Eds.), *Ecosystem health: New goals for environmental management* (pp.57–71). Washington, DC: Island Press.
Sahakian, W. S. (1968). *History of philosophy*. New York: Barnes & Noble Books.
Sahtouris, E. (1989). *Gaia: The human journey from chaos to cosmos*. New York: Pocket Books.
Sankar, Y. (1992). *Education, human values, and ethics: Iimperatives for the information society*. Toronto: Canadian Scholars' Press.
Schaeffer, D. J., & Cox, D. K. (1992). Establishing ecosystem threshold criteria. In R. Costanza, B. G. Norton, & B. D. Haskell (Ed.), *Ecosystem health: New goals for environmental management* (pp. 157–169). Washington, DC: Island Press.
Schweitzer, A. (1949). *Out of my life and thought: An autobiography*. New York: Holt, Rinehart & Winston.
Self, D. J., & Skeel, J. D. (1992). Facilitating healthcare ethics research: Assessment of moral reasoning and moral orientation from a single interview. *Cambridge Quarterly of Healthcare Ethics, 4*, 371–376.
Simon, H. A. (1962). The architecture of complexity. *Proceedings of the American Philosophical Society, 106*, 467–182.
Singer, P. (1975). *Animal liberation*. New York: The New York Review of Books.
Singer, P. (1986). Animals and the Value of Life. In T. Regan (Ed.), *Matters of life and death: New introductory essays in moral philosophy* (2nd ed.) (pp. 338–380). New York: Random House.
Splash C. L. (1993). Economics, ethics, and long-term environmental damages. *Environmental Ethics, 15*, 117–132.
Stewart, T. (1998). The cunning plots of leadership. *Fortune, 138* (5), 165–166.
Stone, C. (1974). *Do trees have standing? Toward legal rights for natural objects*. Los Alamos: William Kaufmann.
Tallmadge, J. (1981). Saying you to the land. *Environmental Ethics, 3*, 351–363.
Thomas, L. (1984). The art of teaching science. In N. R. Comley, D. Hamilton, C. H. Klaus, R. Scholes, & N. Sommers (Eds.), *Fields of writing: Readings across the disciplines* (pp. 559–565). New York: St. Martin's Press.
Tierney, N. L. (1994). *Imagination and ethical ideals: Prospects for a unified philosophy and psychological understanding*. Albany: The State University of New York Press.
Toulmin, S. (1981). The tyranny of principles. *Hastings Center Report, 11*(6), 31–39.
Trusted, J. (1987). *Moral principles and social values*. London: Routledge & Kegan Paul.

REFERENCES

Turner, M. G., Gardner, R. H., & O'Neill, R. V. (1995). Ecological dynamics at broad scales. *BioScience*, Supplement issue, pp. S29-S35.

Ulanowicz, R. E. (1992). Ecosystem health and trophic flow networks. In R. Costanza, B. G. Norton, & B. D. Haskell (Eds.), *Ecosystem health: New goals for environmental management* (pp. 190–206). Washington, DC: Island Press.

U.S. National Commission for the Protection of Subjects of Biomedical and Behavioral Research. (1978). *The Belmont Report: Ethical guidelines for the protection of human subjects*. Washington, DC: DHEW Publications.

Vanier, J. (1998). *Becoming human*. Toronto: Anansi Press.

Waddington, C. (1967). *The ethical animal*. Chicago: University of Chicago Press.

Westra, L. (1994). *An environmental proposal for ethics: The principle of integrity*. Lanham, MD: Rowman & Littlefield.

Whyte, L. L., Wilson, A. G., & Wilson, D. (Eds.). (1969). *Hierarchical structures*. New York: Elsevier.

Wilby, J. (1994). A critique of hierarchy theory. *Systems Practice, 7*, 653–670.

Wilson, E. O. (1978). *On human nature*. Cambridge: Harvard University Press.

Wilson, J. Q. (1993). *The moral sense*. New York: The Free Press.

Wolf, S. A., & Allen, T. F. H. (1995). Recasting alternative agriculture as a management model: The value of adept scaling. *Ecological Economics, 12*, 5–12.

Young, E. W. D. (1989). *Alpha & omega: Ethics at the frontiers of life and death*. Reading: Addison-Wesley.

INDEX

aboriginal peoples, in clear-cutting case study, 80
abortion, 7, 13
animal rights, 15
antibiotic resistance to bacteria, 103–106
applied ethics, 3, 21
archetypal psychology, 116
"The Architecture of Complexity" (Simon), 20

"basic word I-You," 56
beneficence, 5
biodiversity, 24, 32, 88–89
bioethicist, the first, 16
bioethics
  broad scope of, 6, 9
  coining of term, 6, 7
  conflicts involved in, 9–14
  current problems in, 14–17
  defined, 4, 115
  ecological worldview of, 7–9
  environmental concerns on, 7
  framework of scales used in, 40–41
  human experimentation, 4–5
  maternal-fetal issue in, 106–109
  medical advances and, 7–8
  philosophical ethics and morality distinguished from, 3–4, 7
  "principlism" approach to·issues in, 5–6
  Research Ethics, 5
  resource allocations in medicine, 32–33, 96–101
  sterilization of mentally incompetent patients, 109–112
Bioethics in Canada (Roy/Williams/Dickens), 2, 15
Bioethics Service, 2
biophysical systems, 21–22
broader scale, 28–29, 83, 99–100, 102
  See also contextual scale
broad notion of scale, 28–29, 120

Canada: A Model Forest Nation in the Making (House off Commons' Standing Committee on Natural Resources), 71–82
Canadian forest, 69
  See also Christmas Mountains (New Brunswick)
Canadian Forest Service, 78–79
certification system, for consuming forest products, 77
Christmas Mountains (New Brunswick), 36–37, 40, 85–86
clear-cutting case study, 69–96
  Committee considerations in, 71–72

clear-cutting case study (*continued*)
  constituent and contextual scales in, 75–76
  function of forests and, 69–70
  phenomena of concern in, 74–75
  players involved in controversy, 70–71
  recognizing constraints in, 83–96
  recommendations in, 73, 76–82
  relationship of contextual scale to constituents in, 76
  summary of scales in, 83 table
clinical ethics, 4, 5, 14, 28, 40–41
  See also health care
coherence form of moral analysis, 46–47
communal-constituent scale, 75–76, 77
communal-contextual scales, 40, 75–76, 77
communal scale
  in clear-cutting case study, 75, 76, 80–81, 84, 88, 94–95
  ecological issues, 38
  in lake ecosystems issue, 39–40
  in maternal-fetal issue, 108–109
  in resource allocations in medicine issue, 98
communal scale phenomena, 106
community scale, 31–32
complex systems, 20–21, 57, 116
concept of scale, 2, 6
  broadened view of, 28–29, 120
  cosmological nature of ethics, 43–44
  descriptive aspect of. *See* descriptive aspect of scale
  helps in considering complex systems, 116
  "humans within nature" perspective and, 112–113
  human values/choices and, 29
  prescriptive aspect of. *See* prescriptive aspect of scale
  as a source of conflict, 28, 115, 119–120
  used in bioethics, 40–41
  used in ecological systems, 27–28
  is worthwhile for bioethics analysis, 117–118
  See also moral nature of concept of scale; parts and whole framework; scale(s)
concept of sustainability, 58–59
conflict in bioethics, 9–14
  arising from failure to test underlying assumptions, 11–12
  arising from insufficient evidence and experience, 10
  arising from restricted value-focus, 11
  based on partial perceptions, 10–11
  concept of scale and, 28
  on level of ethics, 14
  at level of world-view beliefs, 12–13
  at level of morality, 13–14
conflict over morality, 74
consciousness, intentionality of, 45
constituents
  holon and, 23
  limitations on action of, 66, 99
  on matrix of scales, 38, 39–40
  parts as, 38
  in sterilization of mentally incompetent patients issue, 110
  what is good for the whole is good for the parts approach and, 64–65
constituent scale
  in clear-cutting case study, 73, 75–76, 77, 78, 81, 83, 84, 85–86, 90, 95
  communal scale limiting activity of the, 85
  identifying in narrative approach to scale, 49
  maternal-fetal issue, 108
  mutual consistency and, 67, 93, 106, 117
  pre-eminence of values of contextual scale over, 61–64, 83, 86–87, 120
  in resource allocation in medicine issue, 99, 101
  in social context, 35
  values priority and, 89, 91, 92
  See also local-constituent scale
constrained maximizers, 66
constraints
  in clear-cutting case study, 83–96
  life cycle of a forest insect example, 54–55
  mutual consistency and, 66
  resource allocations in medicine, 98, 99
contextual scale
  in clear-cutting case study, 75–76, 78, 81, 85–86
  higher vs. lower frequency scales and, 54, 55
  holon and, 56
  integrity and good functioning of, 49–50
  in lake ecosystem issue, 39–40
  on matrix of scales, 38
  pre-eminance given to values of, 61–64, 83, 86–87, 120
  prescriptive aspect of scale and, 35, 59
  purpose of, 27, 29
  in resource allocation in medicine case study, 97–98, 98–99
  sterilization of mentally incompetent patients, 110, 111
  See also communal-contextual scales; local-contextual scale

correlative ethics, 37
cosmological nature of ethics, 43–44
"A Critique of Hierarchy Theory" (Wilby), 118
cross-scalar values, 34, 58, 95, 100
cultural and spiritual values, 91

deductive approach to moral analysis, 46
deontology, 53
descriptive aspect of scale, 28, 35, 44, 61, 85, 90, 118
descriptive framework of reality, 44
dialectical thinking, 26
diversity, context and, 63
dream research, 4

ecological dialectic, 26, 27
ecological ethics, 6
ecological scales, 31–32, 87, 102–103, 116
ecological systems, 20, 27–28
ecological values, 70, 91
ecology
    application of concept of scale in, 28
    holistic perspective in, 25–26
    jump from description to prescription in, 44–45
    limits or constraints for high frequency processes in, 54–55
    positivist perspective in, 25–26
    systems theory in, 20
economic values, 88–89, 91
Ecosystem Health and Medicine: Integrating Science Policy and Management conference (1994), 100
environmental ethics, 6, 96
ethics
    applied, 3, 21
    clinical, 4, 5, 14, 28, 40–41
    conflicts and confusions at level of, 14
    constraints to actions and behavior, 54–55
    correlative, 37
    cosmological nature of, 43–44
    intergenerational, 15
    interhuman, 55
    normative, 3, 53
"ethics of intimacy," 52–53
"ethics of strangers," 52–53
ethos, conflicts at the level of, 12–13
evolutionary scale (macro scale), 32
"Experimentation on People" (seminar), 4

fact, values and, 45
federal government, forest management and, 76–77, 78, 88
"first of all, do no harm" principle, 52, 59
forest management
    additional knowledge needed for, 78–79
    federal government responsibility, 76–77
    overseas markets and, 81–82
    sustainable, 77–78
    traditional land use activities and, 80–81
    See also clear-cutting case study
Forest Resource Development Agreement (FRDAs), 79–80, 90–91
forests
    functions of, 69–70
    over-harvesting in, 70
    See also clear-cutting case study; forest management; logging
framework of scales, 37–41

Georgetown University, 5
Global Bioethics (Potter), 7, 16
global health, 101–106
global scale
    change in, 29
    in clear-cutting issue, 74, 75, 82, 84, 86, 95
    global health issue, 103
    from a human perspective, 32
    in matrix of scales, 38
    resource allocations in medicine issue, 98

Hastings Center Report, 112
health care
    maternal-fetal issues in, 106–109
    resource allocation in medicine, 96–101
hierarchical systems
    biophysical, 21–22
    lower *vs.* higher scales and, 22–23
    seen in everyday observations, 21
hierarchy, semantics of, 119
hierarchy theory, 14
    applied to environmental policy-making process, 31
    complex systems, 20–21
    concept of scale from, 2, 28–29
    dialectical thinking and, 26
    on micro and macro scale, 33
    as a systems theory, 118–119
higher-level holons, 23
higher level scales, 22–23, 34, 54–55, 95
"holarchy," 58
holistic perspective on life, 19, 25
holon, 23, 55–57

INDEX

House of Commons Standing Committee on Natural Resources, 94
  See also clear-cutting case study
Hugo, Victor, 14–15
human experimentation
  development of bioethics and, 4–5
  informed consent and, 5
  principles guiding, 5–6
human nature, capacity of foresight and, 59–60
human sociocultural systems, scalar system used in, 33–37, 118
human survival in the natural context, 6

implicate order of reality, 44
individual-case model, 46
individual (local) scale, 31, 38, 39–40
  See also local scale
inductive approach to moral analysis, 46
industry-led approach to forestry, 88–89
informed consent, 5
Institutional Ethics Committee (IEC), 2
insufficient evidence and experience, conflicts based on, 10
intentionality of consciousness, 45
intergenerational ethics, 15
intergenerational scale (meso scale), 31–32
  See also meso scale
interhuman ethics, 55
intermediate community scale, 74
International Convention on Sustainable Forestry, 81–82
International Society for Ecosystem Health and Medicine, 100
intra-scalar debate, 34
is/ought dichotomy, 45–46

Kennedy Institute for Ethics, 5

lake ecosystems, 39–40
"the land ethic," 55
local-constituent scale
  in clear-cutting issue, 75, 76, 84, 95
  in human health and health care issues, 102, 105
local-contextual scale
  in clear-cutting issue, 75, 76, 77, 84, 85
  in human health and health care issues, 102, 104, 105
local scale
  Christmas Mountain logging and, 36–37
  in clear-cutting case study, 72, 74, 75, 77–78, 78, 79, 80, 82, 84, 86–87, 89, 91, 95

explained, 31, 38
global scale and, 29
human health and health care, 49–50, 102, 103, 106
in lake ecosystem issue, 39–40
maintaining integrity of contextual scale and, 83–84
maternal-fetal issue, 107–108
in medical resource allocation case study, 100–101
medical resource allocation issue, 98–99, 101
values of global scale prioritized over, 86
local scale phenomenon, 74, 79
logging, 1–2, 16–17, 34, 36–37, 38–39
  See also clear-cutting case study
long-term perspective to scale
  in clear-cutting case study, 80, 88, 90
  maternal-fetal issue, 108
  priority for contextual scale and, 63
"long-term views," 7
lower-level holons, 23
lower level scales, 22–23, 34, 54, 95
lower scale context, 63
lumber industry, technology used in, 78–79

macroallocation, 32
macro scale, 32, 33, 38, 96, 97–98, 102
  See also global scale; societal scales
magnitude error, 26–27
maternal-fetal issues, 11, 106–109
matrix of scales, 38
media, clear-cutting controversy and, 71
medical experimentation. *See* human experimentation
medicine
  framework of scales in, 40
  resource allocation in, 96–101
  See also health care
mesoallocation, 32
meso scale, 31, 32, 96, 98, 102
metaethics, 3
methicillin, 103
methicillin-resistant *Staphylococcus aureus* (MRSA), 103–104
microallocation, 32
micro scale, 31, 32, 74, 80, 96, 97–98, 101–102
Model Forest Program, 90
moral analysis, coherence form of, 46
moral decision-making, 51–67
  care perspective on, 52–53
  human nature and, 59–60

moral development and, 51–52
psychological underpinning to, 51
scalar, parts, and whole approach to, 60–67
scale and guidance in values priority, 53–59
moral development, progression of stages, 51
morality
  bioethics distinguished from, 3, 7
  conflicts on the level of, 13–14
moral nature of concept of scale
  coherence form of moral analysis, 46–47
  cosmological nature of ethics, 43–44
  is/ought dichotomy, 45–46
  jump from description to prescription and, 44–45
  narrative approach to scale, 47–50
  See also moral decision-making
moral philosophy, hierarchical systems in, 21, 22
mutual consistency, 26, 27, 56, 61, 86, 117

narrative approach to scale, 47–50
National Forest Landscape Management approach, 73
National Forestry Data Base, 78
naturalistic fallacy, 44
natural world
  our relationship with, 6
  responsibilities to, 2
Nazi concentration camps, 4
nonmaleficence, 5
normative ethics, 3, 53
Nuremberg trials, 4

objective description of reality, 25
"observation set," 24

partial perceptions, conflicts based on, 10–11
parts and whole framework, 37
  argument in support of, 120–121
  as a category of conflict, 119
  in clear-cutting case study, 93–94
  moral decision-making, 54–59
  resource allocation in medicine issue, 98–99
  sterilization of mentally incompetent patients issue, 110, 111
  values prioritization and, 116–117
  See also constituent scale; contextual scale
penicillin, 103
phenomenon of concern, 38, 49, 74, 92
philosophical ethics, bioethics distinguished from, 3–4, 7

positivist view, 25
post-positivist view, 25
prescription, jump from descriptive to, 44–45
  See also moral decision-making
prescriptive aspect of scale
  attention to scale and, 50
  bioethics and, 28
  clear-cutting example of, 76, 85, 86, 88, 92
  conceptualization of scale and, 28, 41, 119
  explained, 29–30
  pre-eminence given to values of contextual scale and, 35, 59, 61–64, 66, 108–109
  values prioritization and, 116–117
President's National Commission for the Protection of Human Subjects of Biomedical and Behavioral Research, 5
principle-based normative theories, 53
"principlism" approach to issues in bioethics, 5–6
"The Problem of Scale in Community Resource Management" (Fox), 35–36
public education, forest management and, 77–78

quantitative description of reality, 25
Queen Charlotte Islands, 1–2, 115
Quinlan, Karen Ann, 7

recognized principles, conflicts over, 13–14
regional planning, 112–113
research and development, Canadian Forest Service, 78–79, 89–90
Research Ethics, 5
resource allocation in medicine, 28, 32–33, 96–101
restricted value-focus, conflicts arising from a, 11
Roe v. Wade (1973), 7

scalar approach, examples of
  global health, 101–106
  maternal-fetal issues, 106–109
  resource allocation in medicine, 96–101
  sterilization of mentally incompetent patients, 109–112
  See also clear-cutting case study
scale(s)
  context of the phenomena and, 24
  correlative, 37
  defined, 30
  ecological, 31–32
  knowledge of, 29–30
  "level" vs., 30

scale(s) (continued)
  local, 31
  magnitude error and, 26–27
  multi-dimensional complexity of, 30–31
  narrative approach to, 47–50
  as perspectival, 50
  proposed framework of, 37–41
  resource allocation in medicine example of, 96–101
  in the social context, 33–37
  societal, 32–37
  used in resource allocation issues in medicine, 32–33
  in values priority, 53–55, 53–59
  See also concept of scale
self-interest, 64–65
short-term economic activity, 74
short-term perspective to scale, 7, 74–75
  in clear-cutting case study, 80, 82, 86–87, 89, 91, 93
  priority for contextual scale and, 63
  See also local scale
short-time perspective, 29
social values, 91
societal scales, 32–37, 84, 116
  See also communal scale
space-time scale, 24, 30–31
spatiotemporal scale, 24, 27, 57, 58
  See also long-term perspective to scale; short-term perspective to scale
Staphylococci, 103
Staphylococcus aureus bacteria, 103–104
sterilization of mentally incompetent patients, 109–112
sustainability, concept of, 58–59
sustainable forestry, 73
system, 19
systems
  biophysical, 21–22
  complex, 20–21
  context of the phenomena and, 23–25
  hierarchical, 21
systems theory, 116
  in ecology, 20
  hierarchy theory as a, 118–119

holistic perspective, 19
positivist perspective in, 25–26

technocorporate state, 71
technology, in lumber industry, 78–79
"tragedy of the commons," 8
"The Tyranny of Principles" (Toulmin), 52–53

unexamined assumptions, conflicts arising from, 11–12
U.S. Supreme Court cases, 7
Utilitarian Teleology, 53

values
  in clear-cutting case study, 75–76, 91–92, 94
  economic, 88–89
  emerging from facts, 45
  maintaining the contextual scale, 84–85
  maternal-fetal issue, 108
  reflective separation of fact and, 45–46
  scale, parts, and wholes approach to prioritizing, 60–67
  scale and guidance in prioritizing, 53–59
  scaled, 61
  social and environmental, in logging case study, 91
"values dialogue," 40
views in facts, 52
views in principle, 52

Waddington, C.H., 16
what is good for the whole is good for the part(s) assumption, 64–65
whole is greater than the sum of its parts notion, 26
wholeness of the universe, 44
whole/part approach. See parts and whole framework
withdrawing life-sustaining treatments, 7
women, moral development and, 51–52
worldviews, conflicts occurring at level of, 12–13

Yale Law School, 4

www.ingramcontent.com/pod-product-compliance
Ingram Content Group UK Ltd.
Pitfield, Milton Keynes, MK11 3LW, UK
UKHW021827210426
5322IPUK00003B/67